Meteorology

Atmosphere and Weather

Expanding Science Skills Series

BY

LaVERNE LOGAN AND DON POWERS, Ph.D.

CONSULTANTS: SCHYRLET CAMERON AND CAROLYN CRAIG

COPYRIGHT © 2010 Mark Twain Media, Inc.

ISBN 978-1-58037-527-6

Printing No. CD-404124

Mark Twain Media, Inc., Publishers
Distributed by Carson-Dellosa Publishing LLC

Visit us at www.carsondellosa.com

Table of Contents

Introduction

Meteorology: Atmosphere and Weather is one of the books in Mark Twain Media's new *Expanding Science Skills Series*. The easy-to-follow format of each book facilitates planning for the diverse learning styles and skill levels of middle-school students. The teacher information page provides a quick overview of the lesson to be taught. National science, mathematics, and technology standards, concepts, and science process skills are identified and listed, simplifying lesson preparation. Materials lists for Knowledge Builder activities are included where appropriate. Strategies presented in the lesson planner section provide the teacher with alternative methods of instruction: reading exercises for concept development, hands-on activities to strengthen understanding of concepts, and investigations for inquiry learning. The challenging activities in the extended-learning section provide opportunities for students who excel to expand their learning.

Meteorology: Atmosphere and Weather is written for classroom teachers, parents, and students. This book can be used as a full unit of study or as individual lessons to supplement existing textbooks or curriculum programs. This book can be used as an enhancement to what is being done in the classroom or as a tutorial at home. The procedures and content background are clearly explained in the student information pages and include activities and investigations that can be completed individually or in a group setting. Materials used in the activities are commonly found at home or in the science classroom.

The *Expanding Science Skills Series* is designed to provide students in grades 5 through 8 and beyond with many opportunities to acquire knowledge, learn skills, explore scientific phenomena, and develop attitudes important to becoming scientifically literate. Other books in the series include *Chemistry, Simple Machines, Electricity and Magnetism, Geology, Light and Sound,* and *Astronomy*.

The books in this series promote student knowledge and understanding of science and mathematics concepts through the use of good scientific techniques. The content, activities, and investigations are designed to strengthen scientific literacy skills and are correlated to the National Science Education Standards (NSES), the National Council for Teachers of Mathematics Standards (NCTM), and the Standards for Technological Literacy (STL). Correlations to state, national, and Canadian provincial standards are available at www.carsondellosa.com.

How to Use This Book

The format of *Meteorology: Atmosphere and Weather* is specifically designed to facilitate the planning and teaching of science. Our goal is to provide teachers with strategies and suggestions on how to successfully implement each lesson in the book. Units are divided into two parts: teacher information and student information.

Teacher Information Page

Each unit begins with a Teacher Information page. The purpose is to provide a snapshot of the unit. It is intended to guide the teacher through the developing and implementation of the lessons in the unit of study. The Teacher Information page includes:

- National Standards: The unit is correlated with the National Science Education Standards (NSES), the National Council of Teachers of Mathematics Standards (NCTM), and the Standards for Technological Literacy (STL). Pages 61–65 contain a complete list and description of the National Standards.
- Concepts/Naïve Concepts: The relevant science concepts and the commonly held student misconceptions are listed.
- Science Process Skills: The process skills associated with the unit are explained. Pages 66–68 contain a complete list and description of the Science Process Skills.
- Lesson Planner: The components of the lesson are described: directed reading, assessment, hands-on activities, materials lists of Knowledge Builder activities, and investigation.
- Extension: This activity provides opportunities for students who excel to expand their learning.
- Real World Application: The concept being taught is related to everyday life.

Student Pages

The Student Information pages follow the Teacher Information page. The built-in flexibility of this section accommodates a diversity of learning styles and skill levels. The format allows the teacher to begin the lesson with basic concepts and vocabulary presented in reading exercises and expand to progressively more difficult hands-on activities found on the Knowledge Builder and Inquiry Investigations pages. The Student Information pages include:

1. Student Information: introduces the concepts and essential vocabulary for the lesson in a directed reading exercise.
2. Quick Check: evaluates student comprehension of the information in the directed reading exercise.
3. Knowledge Builder: strengthens student understanding of concepts with hands-on activities.
4. Inquiry Investigation: explores concepts introduced in the directed reading exercise through labs, models, and exploration activities.

Safety Tip: Adult supervision is recommended for all activities, especially those where chemicals, heat sources, electricity, or sharp or breakable objects are used. Safety goggles, gloves, hot pads, and other safety equipment should be used where appropriate.

Unit 1: Historical Perspective
Teacher Information

Topic: Many individuals have contributed to the traditions of the science of meteorology.

Standards:
>   **NSES** Unifying Concepts and Processes, (F), (G)
>   **STL** Technology and Society
>   See **National Standards** section (pages 61–65) for more information on each standard.

Concepts:
- Science and technology have advanced through contributions of many different people, in different cultures, at different times in history.
- Tracing the history of science can show how difficult it was for scientific innovations to break through the accepted ideas of their time to reach the conclusions we currently take for granted.

Naïve Concepts:
- All scientists wear lab coats.
- Scientists are totally absorbed in their research, oblivious to the world around them.
- Ideas and discoveries made by scientists from other cultures and civilizations before modern times are not relevant today.

Science Process Skills:

Students will be **collecting**, **recording**, and **interpreting information** while **developing the vocabulary to communicate** the results of their reading and research. Based on their findings, students will make an **inference** that many individuals have contributed to the traditions of the science of meteorology.

Lesson Planner:
1. Directed Reading: Introduce the concepts and essential vocabulary relating to the history of the science of meteorology using the directed reading exercise found on the Student Information pages.
2. Assessment: Evaluate student comprehension of the information in the directed reading exercise using the quiz located on the Quick Check page.
3. Concept Reinforcement: Strengthen student understanding of concepts with the activities found on the Knowledge Builder page. **Materials Needed:** 2 sheets of 8 1/2" x 11" paper, scissors, glue, colored pencils

Extension: Research the history of the science of meteorology. Create an illustrated time line of scientists and important discoveries.

Real World Application: In 2009, the National Oceanic and Atmospheric Administration and the National Science Foundation sponsored Vortex 2. One hundred scientists and students from 16 universities spent five weeks in the spring chasing and documenting storms in "Tornado Alley."

Unit 1: Historical Perspective
Student Information

Attempts at explaining weather date back to many early civilizations. The early Greeks named the study of weather **meteorology.** Meteorology deals with understanding the forces and causes of weather. Aristotle (384–322 B.C.) used one of the earliest known scientific approaches to weather prediction. His documented work "Meteorologica," detailed his explanation of weather. Aristotle's explanation was based on the interaction of earth, fire, air, and water. He was the first person to describe the water cycle.

Aristotle

Later, one of Aristotle's students, Theophrastus (c. 372–287 B.C.), wrote *The Book of Signs*. This weather textbook is a collection of weather lore and forecast signs. It served as a definitive weather book for almost 2,000 years, until the invention of computers.

Galileo Galilei

Weather instruments were invented to monitor the atmosphere. Galileo Galilei (1564–1642) began the era of recorded meteorological observations with the invention of the earliest form of a thermometer, called a thermoscope. It was a sealed tube with liquid and different weights inside that would rise and fall in a predictable fashion based on how warm or cool the liquid was. This represented the first time that the concepts of hot and cold were freed from Aristotle's concepts of fire and water.

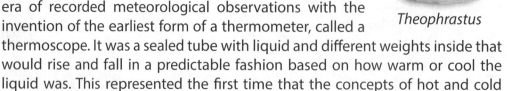

Theophrastus

Daniel Gabriel Fahrenheit

German Daniel Gabriel Fahrenheit (1686–1736) is credited with creating the first temperature scale that could be replicated using a mercury thermometer. He used three different benchmarks to determine at what temperatures water froze (32°F) and boiled (212°F) and what was the average human's body temperature (98.6°F). The Fahrenheit scale was widely used around the world until Anders Celsius developed a scale in which water froze at 0°C and boiled at 100°C. This scale is used in all scientific measurements and in most of the world today.

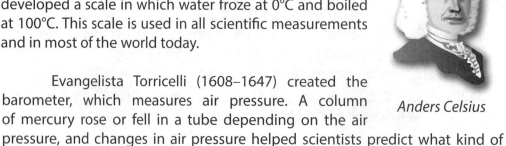

Anders Celsius

Evangelista Torricelli

Evangelista Torricelli (1608–1647) created the barometer, which measures air pressure. A column of mercury rose or fell in a tube depending on the air pressure, and changes in air pressure helped scientists predict what kind of weather was coming. These weather tools were later refined, and we continue to use them to help forecast weather.

Benjamin Franklin

Benjamin Franklin (1706–1790) could be called America's first meteorological scientist. He tried to explain the reasons for various weather-related phenomena. His famous kite experiment in June of 1752 led to the discovery of the static electrical nature of lightning. Franklin also recorded his observations of weather patterns. When clouds obstructed his view of a lunar eclipse, and he discovered that his brother had seen most of the eclipse before the storm got to his town, he hypothesized that the storm was brought in by high southwest winds, even though the surface winds came from the northeast. Benjamin Franklin is often credited with discovering that weather patterns travel from west to east.

All of these tools helped scientists understand the weather, but without a way to communicate with other towns or cities about coming storms, weather prediction remained hit and miss. By 1849, the Smithsonian Institute had established an observation network across the United States. One hundred fifty weather observers were connected by telegraphs. This represented the United States' first attempt at a national weather bureau. By 1860, over 500 telegraph stations were collecting weather data.

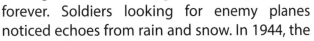
Ulysses S. Grant

The Weather Bureau was established in 1870 when President Ulysses S. Grant signed a joint resolution of Congress authorizing the Secretary of War to establish a national weather service. In 1873, the Weather Bureau issued the first hurricane warning. Weather forecasting became more and more scientific.

Russian-German climatologist Wladimir Köppen (1846–1940) is noted for the Köppen Climate Classification System. In 1884, he published a climatic zone map that showed the seasonal temperature ranges. By 1900, his classification system was created. The Köppen system recognizes five major climate types based on the annual and monthly averages of temperature and precipitation.

Wladimer Köppen

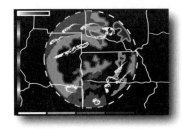

The invention of radar during World War II changed weather prediction forever. Soldiers looking for enemy planes noticed echoes from rain and snow. In 1944, the Great Atlantic Hurricane was seen on radar. Another major weather forecasting advance was the launch of the first weather satellite in 1959. Coupled with the application of computers to forecasting models, more precise forecasts could be made over longer periods of time.

Today, we rely on the National Weather Service to keep us informed of changing weather conditions. The National Weather Service provides Doppler radar images and computer-based weather models to many areas throughout the United States. These tools are vital for improving the accuracy of weather forecasting.

Name: _____   Date: _____

Quick Check

Matching

_____ 1.   meteorology
_____ 2.   barometer
_____ 3.   Fahrenheit scale
_____ 4.   Doppler
_____ 5.   thermoscope

a.   radar tool for weather forecasting
b.   water freezes at 32°F and boils at 212°F
c.   study of weather
d.   used for measuring air pressure
e.   earliest form of a thermometer

Fill in the Blanks

6.   The early Greeks named the study of weather _____.

7.   _____ _____ (1564–1642) constructed the earliest form of a thermometer called a thermoscope.

8.   _____ _____ _____ (1686–1736) is credited for creating a temperature scale that could be replicated using a mercury thermometer.

9.   _____ _____ is often credited with discovering that weather patterns travel from west to east.

10. We rely on the _____ _____ _____ to keep us informed of changing weather conditions.

Multiple Choice

11. In 1884, he published a climatic zone map that showed the seasonal temperature ranges.
     a.   Evangelista Torricelli
     c.   Anders Celsius
     b.   Benjamin Franklin
     d.   Wladimir Köppen

12. He authorized the creation of the Weather Bureau.
     a.   Benjamin Franklin
     c.   Anders Celsius
     b.   Ulysses S. Grant
     d.   Evangelista Torricelli

13. In his documented work "Meteorologica," he was the first to describe the water cycle.
     a.   Gabriel Fahrenheit
     c.   Aristotle
     b.   Benjamin Franklin
     d.   Wladimir Köppen

14. This invention was used first in World War II before it was applied to meteorology.
     a.   radar
     c.   barometer
     b.   hygrometer
     d.   thermometer

Name: _____ Date: _____

Knowledge Builder

Activity: Scientists of Meteorology Mini Book

Directions: Research the history of meteorology. Using your information, create a mini book to display your research.

1. Fold a piece of paper (8 1/2" X 11") as shown below.

2. Cut the paper in half along the fold.

3. Now, fold the two long pieces in half, as shown below (leave one side 1/2" inch shorter than the other side).

4. Next, fold the 1/2" inch tab over the shorter side of each strip.

5. Cut the two strips in half; then cut each half into thirds. This will make 12 small mini books.

6. Use the Student Information and your own reserach to create a scientist mini book. Write the name of a scientist on each of the mini book covers. Lift the matchbook covers and write important information about the scientists.

7. Now, glue the mini books inside a piece of paper (8 1/2" X 11") that has been folded as shown below. There should be 3 rows, with 4 mini books in each row.

8. On the front cover, write a title for your book and put your name on the back cover.

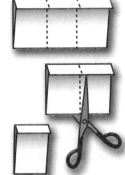

Unit 2: Weather
Teacher Information

Topic: The interaction of air, water, and the sun causes weather.

Standards:
   NSES Unifying Concepts and Processes, (D)
   NCTM Measurement
   See **National Standards** section (pages 61–65) for more information on each standard.

Concepts:
- Weather is considered the conditions of the atmosphere at a particular time and place.
- Weather is due to four atmospheric factors: heat energy, air pressure, winds, and moisture.

Naïve Concepts:
- Weather can be predicted by studying the thickness of the fur on some animals.
- The change in seasons causes the weather to change.

Science Process Skills:

Students will make **observations** about the nature of weather. Students will be **developing vocabulary** related to weather. They will **infer** that weather is due to four atmospheric factors: heat energy, air pressure, winds, and moisture.

Lesson Planner:

1. Directed Reading: Introduce the concepts and essential vocabulary relating to weather using the directed reading exercise found on the Student Information pages.

2. Assessment: Evaluate student comprehension of the information in the directed reading exercise using the quiz located on the Quick Check page.

3. Concept Reinforcement: Strengthen student understanding of concepts with the activities found on the Knowledge Builder page. **Materials Needed:** scissors, 8 1/2" x 11" sheet of white paper, pencil, and colored pencils

Extension: Many people believe that the Farmer's Almanac is a reliable source of weather predictions. Students research the Farmer's Almanac. Using their information, students decide if they agree or disagree with the statement, explain why or why not, and support their claims.

Real World Application: In 1504, Christopher Columbus was having trouble talking the natives of Jamaica into providing his crew with food and supplies. So, after consulting his early "almanac," he told natives if they did not cooperate, he would take the moon away. The lunar eclipse that followed greatly impressed the natives. (Excerpt from *The Handy Weather Answer Book* by Walter Lyons, Page 315).

Unit 2: Weather
Student Information

Weather is the condition of the atmosphere at a particular time and place in a region. Weather impacts what we wear, how we move, our economy, our health, our daily activities, and in short, nearly every aspect of human activity. The interaction of air, water, and the sun causes weather.

The sun causes all weather because it heats the earth unevenly. This uneven heating of the earth causes warm air masses to form at the equator and cold air masses to form at the poles. The warm air masses tend to move toward the poles, while the cold air masses tend to move toward the equator. When these air masses meet and move against each other, clouds and precipitation result. The heat of the sun also causes moisture to rise and form clouds that may bring rain, snow, or thunderstorms. Basically, all the changes in weather come indirectly from the sun.

The wind blows because air has weight. Cold air weighs more than warm air, so the pressure of cold air is greater. As the sun warms the air, the air expands, gets lighter and rises. Heavy, cooler air blows to where the warmer, lighter air was. Usually, wind blows from areas of high pressure to areas of low pressure. Winds can blow very fast if the high pressure area is very close to the low pressure area. Although wind blows from high to low areas of pressure, it doesn't blow in a straight line. This is because the earth is rotating. In the Northern Hemisphere, the spin of the earth causes winds to curve to the right or blow clockwise. In the Southern Hemisphere, winds spin to the left or blow counterclockwise. This is called the **Coriolis Effect**.

The **water cycle**, also known as the hydrologic cycle, plays an important part in our weather. Water may be in one of three basic forms. Water can be in liquid, solid, or gas form. We see water in its liquid form when it rains or when we get a drink from the water fountain. When water freezes,

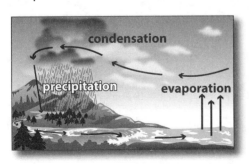

it is in the solid state. Ice and snow are examples of water in the solid state. Water changes from liquid to gas through the process of **evaporation**. It changes from gas to liquid through the process of **condensation**. Air contains a lot of water, in the form of water vapor that is invisible to the naked eye. The moisture in the air is called humidity. Warm air holds more water vapor than cold air. If the air is cold, the water vapor condenses into clouds. When warm air holds a lot of water vapor, it causes our skin to feel moist and sticky.

Severe weather is weather that can potentially cause harm to people and the environment. Hurricanes, tornadoes, thunderstorms, and blizzards are all examples of potentially dangerous storms that may include high winds, hail, flooding, and whiteout conditions.

Quick Check

Matching

_____ 1. water cycle

_____ 2. evaporation

_____ 3. weather

_____ 4. humidity

_____ 5. condensation

a. water changes from liquid to gas

b. moisture in the air

c. hydrologic cycle

d. water changes from gas to liquid

e. condition of the atmosphere at a particular time and place in a region

Fill in the Blanks

6. The interaction of air, water, and the sun causes _____.

7. The _____ causes all weather because it heats the earth unevenly.

8. Cold air weighs more than warm air, so the _____ of cold air is greater.

9. Usually, wind blows from areas of _____ _____ to areas of _____ _____.

10. In the Northern Hemisphere, the spin of the earth causes winds to curve to the right. In the Southern Hemisphere, winds spin to the left. This is called the _____ _____.

Multiple Choice

11. Wind blows because _____.

    a. air has pressure   b. air has weight

    c. the earth spins on its axis   d. of humidity

12. All changes in weather come indirectly from the _____.

    a. wind   b. water

    c. sun   d. Coriolis Effect

13. The spin of the earth in the Northern Hemisphere causes the wind to blow _____.

    a. clockwise   b. counterclockwise

    c. left   d. downward

Name: _____   Date: _____

Knowledge Builder

Activity: Weather Proverbs Booklet

For centuries, people have forecasted weather by clues they gathered from observing plants and animals. For instance, if lightning bugs flashed low to the ground or frogs croaked more often than what was considered normal, this was a sign of rain. Weather sayings or proverbs came from people's observations and are not based on scientific research. However, some proverbs under certain conditions have been proven true by science. For example, "Red sky at night, sailor's delight. Red sky in the morning, sailors take warning."

Directions: Research weather proverbs and make a book to display your favorites.

1. Fold an 8 ½" X 11" sheet of white paper to make eight sections. Number (lightly in pencil) the sections as shown.

    1.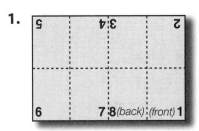

2. Fold the paper in half and cut as shown.

3. Unfold the paper.

    2.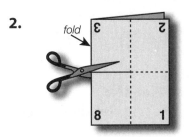

4. Fold the sheet in half, lengthwise, again. Holding one end of the folded page with each hand, lengthwise with the fold at the top, gently push the ends toward the middle. The center sections will move away from each other to form two separate page folds.

5. Fold the sections together so that square 1 is your outer front page and square 8 is the outer back page. Flatten all the folds.

    3.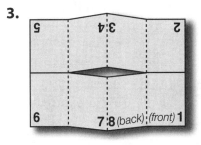

6. Now choose six of your favorite weather proverbs. Write a proverb on each page, illustrate with colored pencils, and tell what the proverb means. Title your booklet *Weather Proverbs*; place your name on the back of the booklet.

4.    **5.**   **6.**

Unit 3: The Sun's Effect on the Atmosphere
Teacher Information

Topic: The sun's effect on earth's atmosphere is the driving force of our weather.

Standards:
>   **NSES** Unifying Concepts and Processes, (D)
>   **NCTM** Number and Operations, Geometry, Measurement, Data Analysis and Probability
>   See **National Standards** section (pages 61–65) for more information on each standard.

Concepts:
- The sun's radiant energy is directly responsible for driving the earth's weather.
- All weather happens in the troposphere.

Naïve Concepts:
- The greenhouse effect is bad and will eventually cause all living things to die.

Science Process Skills:

Students will be formulating hypotheses and conducting experiments. Students will be **using numbers** in the process of **collecting**, **recording**, and **interpreting information** while **developing the vocabulary** to communicate the results of their findings. Based on their findings, students will make an **inference** that the sun plays an important role in our daily weather, seasons, and climate.

Lesson Planner:
1. Directed Reading: Introduce the concepts and essential vocabulary relating to the earth's atmosphere using the directed reading exercise found on the Student Information page.
2. Assessment: Evaluate student comprehension of the information in the directed reading exercise using the quiz located on the Quick Check page.
3. Concept Reinforcement: Strengthen student understanding of concepts with the activities found on the Knowledge Builder page. **Materials Needed:** Activity #1—light source, thermometer; Activity #2—flashlight, white freezer paper or newsprint; Activity #3—globe, flashlight
4. Inquiry Investigation: Explore the relationship between surface color and heat absorption. Divide the class into teams. Instruct each team to complete the Inquiry Investigation pages.

Extension: Students investigate the effects of various earth substances and textures on temperature. They design controlled investigations to determine how various common earth substances such as sand, soil, water, rocks, and grass react to sunlight.

Real World Application: Scientists and governments from around the world have expressed concern over how the composition of the atmosphere is being changed due to human activity. They believe this is affecting global temperature and weather patterns.

Unit 3: The Sun's Effect on the Atmosphere
Student Information

Different layers of air that make up the **atmosphere** surround the earth. The atmosphere extends from Earth's surface to outer space. It contains a mixture of gases. Nitrogen makes up 78 percent while oxygen only makes up 21 percent of Earth's atmosphere. The other 1 percent of Earth's atmosphere includes water vapor and other gases such as argon and carbon dioxide. Earth's atmosphere has been divided into five main layers: troposphere, stratosphere, mesosphere, thermosphere, and exosphere. The **troposphere** is the layer closest to the earth. Seventy-five percent of the atmosphere's gases are located here, and all weather occurs in this layer. The troposphere extends seven to eight miles above the earth, and the majority of our weather data is gathered here. The study of earth's atmosphere and the weather that takes place in it is called **meteorology**. Scientists who study the earth's atmosphere are called **meteorologists**.

The sun's radiant energy plays an important role in our daily weather, seasons, and climate. The sun is responsible for weather phenomena such as blizzards, tornadoes, hurricanes, winds, clouds, fronts, and rainstorms. The effect of the **angle of incidence**, or the angle at which the sun's rays strike the earth, results in the uneven heating of the earth's surface. The uneven heating of the earth's surface by the sun is the catalyst for much of this weather. The angle of incidence is influenced by the rotation of the earth as it revolves around the sun.

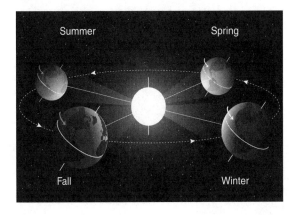

The heating and cooling of the earth and its atmosphere is a natural part of weather and climate. Water heats and cools more slowly than rocks, sand, and soils. Polar water remains cold throughout the year, while tropical water retains large amounts of heat throughout the year, despite the effects of seasons. Temperatures in land regions, on the other hand, are much more influenced by the earth's seasons. The line between air warmed and cooled by tropical and polar waters and air over land regions is a critical element to understanding our weather.

Weather is the condition of the atmosphere at a particular time and place in a region. The interaction of air, water, and the sun causes weather. Air masses moving across the earth cause different weather phenomena to occur. These air masses are warmed when the sun shines on and heats the earth. The equator is always heated more by the sun than the North and South Poles. This uneven heating of the earth causes warm air masses to form at the equator and cold air masses to form at the poles. The warm air masses tend to move toward the poles, while the cold air masses tend to move toward the equator. When these air masses meet and move against each other, clouds and precipitation result.

Quick Check

Matching

_____ 1.   atmosphere

_____ 2.   meteorology

_____ 3.   troposphere

_____ 4.   weather

_____ 5.   angle of incidence

a.   different layers of air that surround the earth

b.   the layer of the atmosphere closest to the earth

c.   the angle at which the sun's rays strike the earth

d.   study of earth's atmosphere

e.   condition of the atmosphere at a particular time and place in a region

Fill in the Blanks

6.   The sun's _____ _____ plays an important role in our daily weather, seasons, and climate.

7.   The angle of incidence is influenced by the _____ of the earth as it _____ around the sun.

8.   Earth's _____ has been divided into five main layers.

9.   The _____ is responsible for weather phenomena such as blizzards, tornadoes, hurricanes, winds, clouds, fronts, and rainstorms.

10. The _____ is always heated more by the sun than the North and South Poles.

Multiple Choice

11. The troposphere extends up _____ to _____ miles.
     a.   three to five
     b.   seven to eight
     c.   nine to ten
     d.   two to three

12. Nitrogen makes up _____ percent of Earth's atmosphere.
     a.   78
     b.   21
     c.   1
     d.   92

13. Seventy-five percent of the atmosphere's gases are located here.
     a.   stratosphere
     b.   troposphere
     c.   mesosphere
     d.   thermosphere

Name: _____   Date: _____

Knowledge Builder

Activity #1: Lamp and Thermometer Demonstration

Directions: Arrange three light sources at angles so they all shine on thermometers from directly overhead, 45 degrees, and 90 degrees. Be sure that the distance between the lamps and the thermometers is equal. Record the

Overhead   45°   90°

temperature of each thermometer in Celsius in the data table below. Turn the light source on. Record the temperature for each thermometer every 15 minutes for one hour in the data table below.

Time	Temperature in Celsius		
	Directly Overhead	**45 Degrees**	**90 Degrees**
Starting Temperature			
15 minutes			
30 minutes			
45 minutes			
60 minutes			

Conclusion: How does the angle at which light from the sun strikes the earth affect temperature?

_____

Activity #2: Flashlight and Paper

Directions: Place white paper flat on a table. Measure a height of 50 centimeters. From this height, shine the flashlight directly overhead onto the paper (lens of the flashlight should be parallel to the table top). Have one person

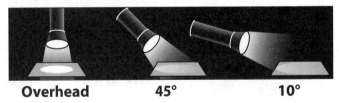

Overhead   45°   10°

trace and label the outline of the light pattern from this angle. Next, while holding the distance constant; rotate the flashlight 45 degrees to one side. Again, trace and label the outline of the light pattern. Lastly, rotate the flashlight to 10 degrees (again, keep the distance constant), and trace and label the outline of the light pattern. Calculate and record the area of each of the three patterns in the data table below (estimate in square centimeters if necessary).

Area (sq. cm)		
Directly Overhead	**45 Degrees**	**10 Degrees**

Conclusion: What implications for our weather does your data indicate? _____

_____

Name: _____   Date: _____

Knowledge Builder

Activity #3: Angle of Incidence

Directions: Using a globe, hold the flashlight 10 cm away from the equator. With the rays striking the equator line directly, carefully observe the point at which the sun's rays strike the earth. Record a drawing of the path of the sun's rays in the data table below. Next, holding the light steady, raise it straight up so that the light shines roughly on the latitudes of the United States. Observe and record the path of the rays as before. Move the flashlight up to the North Pole and observe and record the path of the rays again.

Sun's Rays		
Directly at Equator	**United States**	**North Pole**

Conclusions

1. The earth is not always the same distance from the sun during the year. Explain.
    _____
    _____

2. The sun's rays do not all travel the same distance to reach the earth. Explain.
    _____
    _____

3. In what ways would our weather be affected if the earth were not tilted?
    _____
    _____

4. In what ways would our weather and world be affected if the earth did not rotate?
    _____
    _____

Name: _____ Date: _____

Inquiry Investigation: Surface Color and Temperature

Concepts:
- Surface color and texture affect heat absorption.
- The color and texture of earth materials impact temperatures when exposed to sunlight.

Purpose: Does the color of a surface affect temperature?

Hypothesis: Write a sentence that predicts what you think will happen in the experiment. Your hypothesis should be clearly written. It should answer the question stated in the purpose.

Hypothesis: _____

_____

Procedure: Carry out the investigation. This includes gathering the materials, following the step-by-step directions, and recording the data.

Materials:
various colors of construction paper (black, white, brown, green, and yellow)
tape　　　　　　　　　thermometers　　　　　　　　　scissors
stapler　　　　　　　　sunny location or light source

Experiment:
Step 1: Construct a paper pocket from each sheet of construction paper.
Step 2: Insert a thermometer into the open end of each pocket. Tape the top of the pocket closed with the thermometer still inside, while allowing the thermometer to be slipped out to be read.
Step 3: Read the thermometer and record the temperature (°C) in the data table below. Place the thermometer inside the black pocket.
Step 4: Repeat Step 3 for each of the other colored pockets.
Step 5: Place the pockets and thermometers in direct sunlight or under a light source.
Step 6: Record the temperatures in the data table below at 15-minute intervals for a period of one hour.

Results: Complete the data table below. Record the temperature for the times in the data table.

Paper Color	Temperature (°C)			
	15 min.	30 min.	45 min.	60 min.
black				
white				
brown				
yellow				
green				

Name: _____ Date: _____

Analysis: Study the results of your experiment. Decide what the data means. This information can then be used to help you draw a conclusion about what you learned in your investigation.

Use your data to create a line graph to track changes over time in temperature. Plot the temperature on the *y*-axis. Use colored pencils to represent the colors of the pockets on the graph. Plot the time on the *x*-axis.

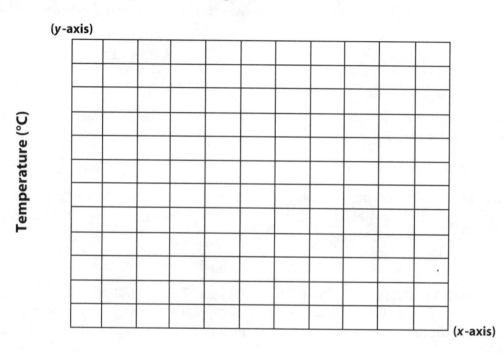

Change of Temperature Over Time

(*y*-axis)

Temperature (°C)

(*x*-axis)

Time (min)

Conclusion: Write a brief description of what happened in the experiment and whether or not your hypothesis was supported by the data.

_____

_____

_____

_____

_____

_____

_____

Unit 4: Air Pressure
Teacher Information

Topic: Air is a key factor to our weather.

Standards:
   NSES Unifying Concepts and Processes, (D)
   NCTM Measurement and Data Analysis and Probability
   STL Technology and Society
   See **National Standards** section (pages 61–65) for more information on each standard.

Concepts:
- Air has certain properties: takes up space, has mass, and is a fluid.
- Air has mass and volume; therefore it exerts pressure.
- The symbols **H** and **L** on a weather map represent areas of high and low pressure. The isobars represent areas of equal pressure.

Naïve Concepts:
- Air neither has mass nor can it occupy space.
- The **H** on weather maps stands for hot temperatures and the **L** means cold weather.
- Isobars on weather maps represent wind speed or temperature.

Science Process Skills:

Students will **compare** and **contrast** weather conditions associated with areas of high and low pressure. Students will be **communicating** and **developing vocabulary**. Students will be **using numbers** during the process of **collecting**, **recording**, **analyzing**, and **interpreting** data. Students will **infer** that since air has mass and volume, it must therefore exert pressure.

Lesson Planner:
1. <u>Directed Reading</u>: Introduce the concepts and essential vocabulary relating to air pressure using the directed reading exercise found on the Student Information pages.
2. <u>Assessment</u>: Evaluate student comprehension of the information in the directed reading exercise using the quiz located on the Quick Check page.
3. <u>Concept Reinforcement</u>: Strengthen student understanding of concepts with the activities found on the Knowledge Builder page. **Materials Needed:** Activity #1—hard-boiled egg, newspaper, matches, glass milk bottle; Activity #2—clean small tin can (be cautious of sharp edges), large balloon, rubber band, straw, 3" x 5" index card, pen or marker, aneroid barometer

Extension: Students make observations about properties of air. Distribute bubble wrap sheets to students. Instruct students to observe the bubbles and make observations they can defend regarding the properties of air.

Real World Application: Air pressure can affect an athlete's performance. There is less resistance from the atmosphere at high altitudes, which means that an athlete has the potential to jump higher or run faster. However, the higher the altitude, the less oxygen is available, which means the body has to work harder, which often leads to slower running times.

Unit 4: Air Pressure
Student Information

Air pressure is also known as atmospheric pressure or barometric pressure. It is a measure of the weight of air pressing down on a given area of Earth's surface. Air pressure is caused by the weight of air from the top of the atmosphere pressing down on the layers of air below. The layers press down on each other because gravity pulls the air molecules down.

Air has mass and volume. As a result, it exerts pressure. Air is also considered to be a **fluid** as it has the ability to take the shape of its container and to flow. Meteorologists identify air masses of low pressure and high pressure as they track weather patterns and prepare forecasts. Meteorologists use an instrument called a **barometer** to measure changes in air pressure. Today, aneroid barometers have largely replaced mercury barometers due to safety concerns about mercury; however, the data is still reported in units of inches, which came from measuring the height of the column of mercury.

The weather conditions associated with air masses of varying pressures are different. Regions of sinking cool air are called **high-pressure systems**, or anticyclones. In a high-pressure system, the winds rotate clockwise. Regions of rising warm moist air are called **low-pressure systems**, depressions, or cyclones. In a low-pressure system, the winds rotate counterclockwise. If the pressure is very low, these spiraling winds may reach storm or hurricane force. Areas of high pressure are usually associated with fair weather, whereas areas of low pressure are commonly associated with stormy weather. On weather maps, these air masses are labeled with a large **L** for low pressure and **H** for high pressure.

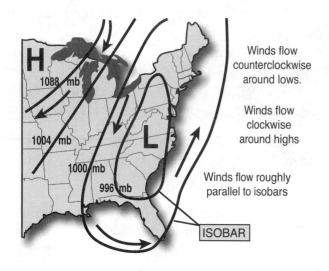

Winds flow counterclockwise around lows.

Winds flow clockwise around highs

Winds flow roughly parallel to isobars

ISOBAR

Meteorologists identify air masses as high and low pressure by collecting barometric pressure data from weather reporting stations across the United States. After the barometric pressure data is plotted on a map, lines are drawn to connect areas of equal pressure. These lines are referred to as **isobars**, or lines of equal pressure. Barometric pressure decreases as one moves to the center of the low-pressure area; barometric pressure increases as one moves to the center of a high-pressure area.

Name: _____ Date: _____

Quick Check

Matching

_____ 1. low-pressure systems
_____ 2. air pressure
_____ 3. barometer
_____ 4. isobars

_____ 5. high-pressure systems

a. used to measure changes in air pressure

b. regions of sinking cool air

c. lines of equal pressure on a weather map

d. known as atmospheric pressure or barometric pressure

e. regions of rising warm moist air

Fill in the Blanks

6. Air pressure is caused by the _____ of air from the top of the atmosphere pressing down on the layers of air below.

7. In a _____ _____ _____, the winds rotate counterclockwise.

8. Areas of high pressure are usually associated with _____ _____.

9. Areas of low pressure are commonly associated with _____ _____.

10. On weather maps, air masses are labeled with a large _____ for low pressure and _____ for high pressure.

Multiple Choice

11. What is another name for a high-pressure system?
    a. anticyclone
    b. depression
    c. cyclone
    d. hurricane

12. Where do meteorologists get their barometric pressure data?
    a. clouds
    b. weather reporting stations
    c. barometers
    d. telephone books

13. What is air considered to be because it can take the shape of its container and it can flow?
    a. solid
    b. volume
    c. fluid
    d. area

Name: _____ Date: _____

Knowledge Builder

Activity #1: Egg in a Bottle

Directions: CAUTION: This experiment should be conducted under adult supervision. Remove the shell from a hard boiled egg. Dampen the egg with water. Now, drop a small piece of burning newspaper into a glass milk bottle. Place the egg on the opening of the bottle.

What happened to the egg? _____

_____

Conclusion: What caused the egg to fall into the bottle?

_____

_____

To remove the egg from the bottle, hold the bottle upside down with the small end of the egg in the jar neck. Tilt the bottom of the bottle down until there is a small opening between the neck of the bottle and the egg. Blow hard into the bottle, making a closed seal with your mouth. Before you remove your mouth, tilt the bottle upside down. The egg should drop out.

Activity #2: Construct a Barometer

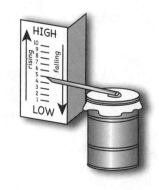

Directions: Cut a large balloon in half. Stretch the balloon over the top of a clean tin can; stretch it so it is tight. Have a partner hold the balloon tight while you place a rubber band over the balloon to secure it in place. Cut a straw 6 cm in length. Cut one end at an angle to create a pointed edge. Glue the non-cut end of the straw to the middle of the stretched balloon on the top of the can. Fold a 3" x 5" index card in half. Stand it on edge next to the straw pointer, and mark the point at which the straw pointer touches the card; label this line 5. Measure every half-centimeter above and below and mark 6, 7, 8, 9, 10 going up, and 4, 3, 2, 1, going down. Write the word *high* near the number 10 and the word *low* near the number 1. Along one side, write the word *rising* and an arrow pointing up. Along the other side, write the word *falling* and an arrow pointing down. (See diagram.)

      Now you are ready to begin collecting data. Check the barometer in the morning and the afternoon; record the number by the straw pointer as well as falling or rising as it goes up and down. Compare your rising and falling data with that of data collected from an aneroid barometer provided by your teacher.

Conclusion: What types of weather conditions are present during times when the barometric pressure is falling and low? Rising and high? _____

_____

Name: _____   Date: _____

Knowledge Builder

Activity #3: Weather Map

Directions: A meteorologist has begun charting the barometric pressure by writing the barometric pressure on the isobars. She has labeled the isobars in millibars.

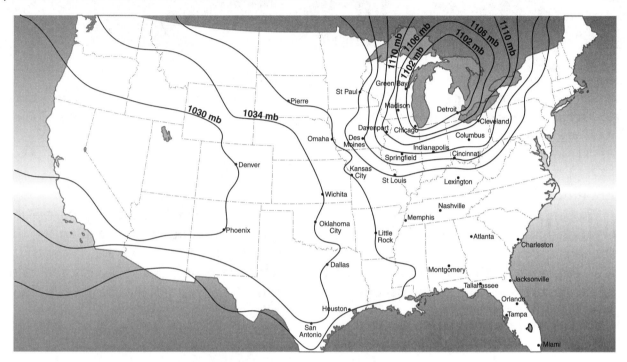

1. Finish labeling the isobars that run through the following cities:
    A. Springfield, IL, and Indianapolis, IN
    B. Des Moines, IA, and Cincinnati, OH
    C. St. Paul, MN, and St. Louis, MO
    D. Pierre, SD; Omaha, NE; Kansas City, MO; Little Rock, AR; and Houston, TX

2. Describe patterns you notice in the barometric pressure readings in the upper Midwest and the West-Southwest. _____

    _____

3. Identify an area of low pressure and an area of high pressure; write "L" for low and "H" for high on the map.

4. Based upon the barometric pressure, what type of weather is likely occurring in Michigan? In the West-Southwest? Explain your reasoning._____

    _____

    _____

Unit 5: The Water Cycle
Teacher Information

Topic: The water cycle plays an important role in our weather.

Concepts:
- The water cycle describes the continuous exchange of water between land, bodies of water, and the atmosphere.

Naïve Concepts:
- The same water cycles in the same part of the earth over and over again.

Science Process Skills:

Students will **develop** vocabulary relating to the water cycle. Students will be making **observations**, **collecting**, **recording**, and **interpreting information** while **developing the vocabulary to communicate** the results of their findings. Based on their findings, students will make an **inference** that water evaporates from the earth's surface, rises and cools as it moves to higher elevations, condenses as rain or snow, and falls to the surface where it collects in lakes, oceans, soil, and rocks underground.

Lesson Planner:
1. <u>Directed Reading</u>: Introduce the concepts and essential vocabulary relating to the water cycle using the directed reading exercise found on the Student Information page.
2. <u>Assessment</u>: Evaluate student comprehension of the information in the directed reading exercise using the quiz located on the Quick Check page.
3. <u>Concept Reinforcement</u>: Strengthen student understanding of concepts with the activities found on the Knowledge Builder page. **Materials Needed:** Activity #1—hot glue gun, glue stick, black felt, cardboard, snowy day; Activity #2—resealable plastic bag, red food coloring, water, small plastic cup, tape

Extension: Students investigate whether the color of water has any effect on the evaporation/condensation process.

Real World Application: In the 1930s, rolling dust storms swept across the Great Plains with devastating effects on the people and land. Drought, combined with prior mistreatment of the land, led to one of the greatest natural disasters in the United States, the Dust Bowl.

Unit 5: The Water Cycle
Student Information

The **water cycle**, or hydrologic cycle, explains how water exists on the earth. To understand the water cycle, you must first understand that the amount of water on earth is constant. The amount of water on the earth has remained the same since its formation. The state of matter—solid, liquid, or gas (vapor) in which the water exists can and does change. The atmosphere can hold a varying amount of water, either as water vapor or as liquid or solid water in clouds or some form of precipitation. Much of the earth's water is contained by the oceans, lakes, ponds, and other bodies of water. This is the water that is most commonly visible to you.

Water can also be contained in the ground in the form of **groundwater**. This groundwater may be hidden in underground aquifers (an underground layer of rock that holds fresh water) or in-ground. Water on the earth can also be contained in plants and animals. About three-fourths of the human body is water contained in cells, tissues, and organs. Water in the solid state can also exist as glaciers and icebergs. Most of the water that comes to the ground from the atmosphere arrives in some form of precipitation, such as rain, snow, dew, etc. The water that lands on the earth's surface may be either absorbed into the ground and become part of the groundwater system or it might run into a lake, stream, pond, or some other body of water on Earth's surface.

The water on the earth and in the atmosphere continuously moves throughout the earth and its atmosphere in the repeated process of evaporation, condensation, and precipitation.

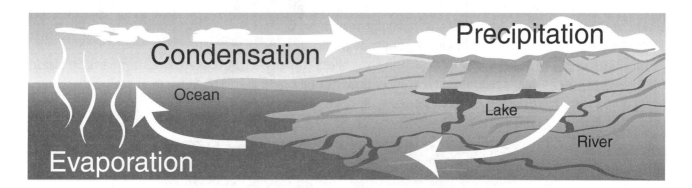

The water in the atmosphere usually gets into the atmosphere by way of evaporation. **Evaporation** is the moving of water from a liquid state to a gaseous state below its boiling point. Evaporation occurs when the sun heats up water in rivers, lakes, or the ocean and turns it into vapor or steam. As water evaporates, its molecules move more quickly than when it was a liquid. The energy absorbed is stored in vaporous molecules as **latent** (meaning hidden) **heat**. During evaporation, molecules of water in the liquid state move or vibrate with enough energy to be released from other water molecules and travel into the atmosphere. The temperature of the air determines the amount of water vapor the air can hold. Warmer air has a higher capacity to hold water vapor than cooler air.

A key phase in the water cycle is **condensation**, the movement of water from a gaseous state to a liquid state. Condensation can be thought of as the opposite of evaporation. Condensation occurs as air with water vapor in it cools. This is why water droplets form on cold glasses or cans. The air around the beverage is cooling to the temperature of the beverage, pulling the water vapor out of the air. Condensation is the basis by which clouds are formed, both in the sky and near the ground as fog.

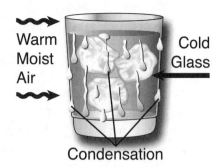

Warm Moist Air

Cold Glass

Condensation

Another key phase of the water cycle is precipitation. The earth's water cycle is responsible for all the precipitation on the earth. Precipitation is arguably the most prominent feature of the water cycle in our weather. It is directly observable and affects our daily lives. **Precipitation** occurs when water droplets in liquid and solid form become so large and heavy they can no longer stay in the earth's atmosphere. They are forced to obey the pull of gravity, and they fall as precipitation.

The most common forms of precipitation include rain and snow. Rain, a liquid water, is usually measured in a calibrated container called a rain gauge with straight sides and a flat bottom. When the air is cold enough, water vapor immediately turns to solid snow, made up of six-sided crystals.

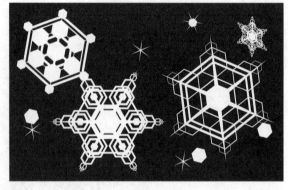

Less common forms of precipitation include sleet, freezing rain, and hail. Sleet is partially frozen water that reaches the ground in winter, but may melt before hitting the ground in summer. When ice pellets form and get tossed around in a high cloud until several successive layers of water freeze, they get heavy enough to fall as hail. Hail pellets vary from the size of rice grains to softballs or even small melons.

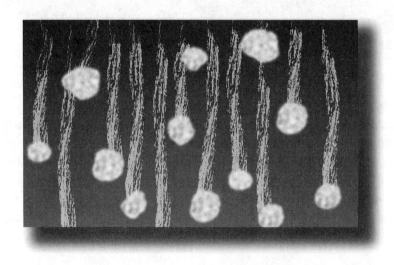

Name: _____ Date: _____

Quick Check

Matching

_____ 1. groundwater

a.   the moving of water from a liquid state to a gaseous state

_____ 2. condensation

b.   occurs when water in liquid and solid form can no longer stay in the earth's atmosphere

_____ 3. water cycle

c.   water contained in the ground

_____ 4. evaporation

d.   the movement of water from a gaseous state to a liquid state

_____ 5. precipitation

e.   an exchange of water between land, bodies of water, and the atmosphere

Fill in the Blanks

6.   _____ can be thought of as the opposite of evaporation.

7.   Condensation is the basis by which _____ are formed, both in the sky and near the ground as fog.

8.   Evaporation occurs when the _____ heats up water in rivers, lakes, or the ocean and turns it into vapor or steam.

9.   The earth's water cycle is responsible for all the _____ on the earth.

10. The water in the atmosphere usually gets into the atmosphere by way of _____.

Multiple Choice

11. What is energy absorbed and stored in vaporous molecules called?

a.   precipitation   b.   cool heat

c.   latent heat   d.   condensation

12. Which of the following is NOT a process in the water cycle?

a.   groundwater   b.   evaporation

c.   condensation   d.   precipitation

13. An underground layer of rock that holds fresh water is called _____.

a.   latent heat   b.   sleet

c.   the hydrologic cycle   d.   an aquifer

Name: _____ Date: _____

Knowledge Builder

Activity #1: Snowflakes

Directions: Construct a snowflake catcher by gluing a piece of black felt to a stiff piece of cardboard. Place the snowflake catcher in the freezer, so when the snowflakes are "caught," they will not melt immediately. Go outdoors to catch and observe snowflakes. Use a hand lens to compare and contrast the snowflakes' size and shape. Use reference materials* to identify the types of crystal formations that were observed.

*A complete classification of snowflakes can be found at "Winter's Story: Basic Snowflake Identification Field Data." NASA. <www.nasa.gov/pdf/182187main_BasicSnowflakeFieldData. pdf>

Activity #2: Water Cycle in a Bag

Directions: Fill a small cup about one-half full of water. Add a drop or two of red food coloring to the water. Place the cup upright in one corner of the resealable plastic bag as you hold the opposite corner of the bag up. Seal the bag. Tape the bag upright in a location that receives daily sunlight. Observe the bag daily for a week and draw any changes you observe in the data table below.

Day 1	Day 2	Day 3	Day 4	Day 5

Observation: What happened in the bag? Why? _____

_____

_____

Conclusion: How is your water cycle in a bag similar to the one found in nature? _____

_____

_____

_____

Unit 6: Wind
Teacher Information

Topic: Wind speed and direction affect weather systems.

Standards:
   NSES Unifying Concepts and Processes, (D)
   NCTM Geometry, Measurement, and Data Analysis and Probability
   STL Abilities for a Technological World
   See **National Standards** section (pages 61–65) for more information on each standard.

Concepts:
   • Winds are produced by the uneven heating of Earth's surface and the resulting rise and fall of differentially heated air masses

Naïve Concepts:
   • Clouds block wind and slow it down.
   • Cold temperatures produce fast winds.

Science Process Skills:
   Students will make **observations** about the type of weather associated with different wind speeds and directions. Students will be **communicating** and **developing vocabulary**. Students will be **using numbers** during the process of **collecting**, **recording**, **analyzing**, and **interpreting** data. They will **infer** that wind speed and direction affect weather systems.

Lesson Planner:
   1. <u>Directed Reading</u>: Introduce the concepts and essential vocabulary relating to wind using the directed reading exercise on the Student Information pages.
   2. <u>Assessment</u>: Evaluate student comprehension of the information in the directed reading exercise using the quiz located on the Quick Check page.
   3. <u>Concept Reinforcement</u>: Strengthen student understanding of concepts with the activities found on the Knowledge Builder page. **Materials Needed:** Activity #2—sharpened pencil with eraser, straight pin, tag board, scissors, 6 oz. Styrofoam cup

Extension: Students research wind chill and how calculations to determine wind chill have changed recently.

Real World Application: Many local wind systems have their own names. Chinooks are warm winds originating off the Pacific coast. The winds cool as they climb the western slopes of the Rocky Mountains, and then rapidly warm as they drop down the eastern side of the mountains. Strong, dry winds that originate off the Pacific coast and blow over Southern California from fall to winter are called the Santa Ana winds.

Unit 6: Wind
Student Information

Wind, air in motion, is a result of the uneven heating of the earth's surface by the sun. As the sun warms the Earth's surface, the atmosphere warms too. Warm air rises. Then cool air moves in and replaces the rising warm air. This movement of air is what makes the wind blow. Meteorologists are primarily concerned with two facets of wind as they prepare and deliver forecasts—speed and direction.

The wind blows because air has weight. Usually wind blows from areas of high pressure to areas of low pressure. Cold air weighs more than warm air, so the pressure of cold air is greater. As the sun warms the air, the air expands, gets lighter, and rises. Winds blow in a counterclockwise direction in regions of rising warm moist air and are called **low-pressure systems**. Clouds, rain, or snow and strong winds often occur in these regions. Heavy cooler air blows to where the warmer, lighter air was. Winds blow in a clockwise direction in regions of sinking cool air and are called **high-pressure systems**. Clear skies and fair weather usually occur in these regions. Winds can blow very fast if the high pressure area is very close to the low pressure area.

Local winds: some local winds reflect the fact that land heats faster than water, but water retains its heat longer. Cooler moist air from over the water slides beneath the rising air, creating a **sea breeze** on an ocean beach. At night, the water stays warm after the land has cooled, resulting in an **offshore breeze**. The name of a breeze always tells the direction from which the wind is blowing. A southwest breeze blows from the southwest toward the northeast. Some "local "winds can be quite regional and seasonal in nature, like the warm and wet monsoon winds that bring heavy rain to Southeast Asia when they blow from ocean to land.

Global winds: some winds blow over long distances from specific directions. Solar energy pours more efficiently into the atmosphere at the equator where it enters from directly overhead. In general, warm tropical air tends to flow toward the cool poles, while polar air slides toward the equator. Earth's air masses break up into cells that result in certain consistent wind patterns.

- **Doldrums:** mostly very calm air in a band over the equator.

- **Trade winds:** warm, steady winds blow back toward the equator in usually clear skies.

- **Prevailing westerlies:** cool air, usually moving quickly toward the poles from west to east in both hemispheres.

- **Polar easterlies:** cold, fairly weak winds blowing from east to west.

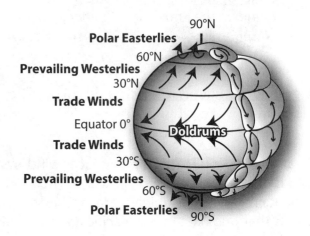

Name: _____ Date: _____

Quick Check

Matching

_____ 1. high-pressure systems
_____ 2. wind
_____ 3. polar easterlies
_____ 4. doldrums
_____ 5. low-pressure systems

a. cold, fairly weak winds blowing from east to west
b. regions of rising warm moist air
c. air in motion
d. regions of sinking cool air
e. mostly very calm air in a band over the equator

Fill in the Blanks

6. _____ _____ pours more efficiently into the atmosphere at the equator where it enters from directly overhead.

7. In general, warm tropical air tends to flow toward the cool _____, while polar air slides toward the _____.

8. At night, the water stays warm after the land has cooled, resulting in an _____ _____.

9. Cooler moist air from over the water slides beneath the rising air, creating a _____ _____ on an ocean beach.

10. The name of a breeze always tells the _____ from which the wind is blowing.

Multiple Choice

11. These warm, steady winds blow back toward the equator in usually clear skies.
    a. prevailing westerlies
    b. polar easterlies
    c. doldrums
    d. trade winds

12. Which way does a southwest breeze blow?
    a. from the southwest toward the northeast
    b. from the southeast toward the northwest
    c. from the northwest toward the southwest
    d. from the southwest toward the northwest

13. These winds contain cool air, usually moving quickly toward the poles from west to east in both hemispheres.
    a. doldrums
    b. prevailing westerlies
    c. trade winds
    d. polar easterlies

Name: _____ Date: _____

Knowledge Builder

Activity #1: Wind Speed

Directions: Wind speed and direction are key weather data. Record and measure the wind speed and direction every day for four weeks. Use the chart of visual wind speed indicators below to measure wind speed. Create a data table to record your data

Observation	Description	Miles per Hour
smoke goes straight up	calm	1–3
smoke moves, but wind vane does not	light breeze	4–7
leaves rustle; wind felt on face	gentle breeze	8–12
leaves and small twigs move constantly; wind extends light flag	moderate breeze	13–18
small trees sway; small waves crest on lakes	fresh breeze	19–24
large branches move constantly; wires on electric poles hum	strong breeze	25–31
large trees sway; walking against wind is inconvenient	moderate gale	32–38
twigs break off large trees; walking against wind is difficult	fresh gale	39–46
branches break off trees; loose bricks blow off chimneys; shingles blow off buildings	strong gale	47–54
trees snap or are uprooted; considerable damage to buildings is possible	whole gale	55–63
widespread damage to buildings	storm	64–75

Activity #2: Wind Vane

Directions: Construct this wind vane to help determine wind direction. Cut out the pointer and fin on tag board. Tape the pointer and the fin at opposite ends of a straw (see diagram). Invert a 16-oz. Styrofoam cup and carefully bore a hole in the center with a sharpened pencil. Push the pencil down until it is level with the bottom of the inverted cup. The pencil should fit snugly in the hole that was created. Push a straight pin through the straw and mount it on the top of the pencil by pushing the pin into the eraser. Balance the straw with the pointer and fins so it is level and freely spins around in the wind. You may need to adjust the centering of the pin after testing it outdoors. The wind vane will point to the direction from which the wind is blowing.

Conclusion: Why is the daily collection of wind speed and direction important?

_____

_____

Unit 7: Clouds
Teacher Information

Topic: Clouds, formed by condensation of water vapor, affect weather and climate.

Standards:
> **NSES** Unifying Concepts and Processes, (D)
> **NCTM** Geometry, Measurement, and Data Analysis and Probability
> **STL** Abilities for a Technological World
> See **National Standards** section (pages 61–65) for more information on each standard.

Concepts:
- A cloud is primarily tiny water droplets and/or thin ice crystals.
- Certain types of clouds are associated with certain types of weather.

Naïve Concepts:
- A cloud is made of water vapor.
- Clouds in the sky foretell rain.

Science Process Skills:

Students will make **observations** about clouds and the type of weather associated with the different cloud formations. Students will be **communicating** and **developing vocabulary**. Students will be **using numbers** during the process of **collecting**, **recording**, **analyzing**, and **interpreting** data. They will **infer** that clouds are formed by condensation of water vapor and clouds affect weather and climate.

Lesson Planner:
1. Directed Reading: Introduce the concepts and essential vocabulary relating to clouds using the directed reading exercise found on the Student Information page.
2. Assessment: Evaluate student comprehension of the information in the directed reading exercise using the quiz located on the Quick Check page.
3. Concept Reinforcement: Strengthen student understanding of concepts with the activities found on the Knowledge Builder page. **Materials Needed:** Activity #1—clean soup can, thermometer, water, ice, glass stirring rod; Activity #2—2 ounces of water, graduated cylinder, plastic 2-liter soda bottle with cap, matches; Activity #3—thermometer, psychrometer, barometer, wind vane, cloud chart; Activity #4—two thermometers, hollow shoestring, water, manila folder

Extension: Students investigate the relationship between the time of year (season) and moisture levels in their home.

Real World Application: Perspiring or sweating helps our bodies cool down. The relative humidity determines the rate at which the water can evaporate from the skin. When the air is full of moisture, it is harder for the air to absorb the sweat from our skin. The result is that we feel hot and sticky on summer days with high humidity

Unit 7: Clouds
Student Information

Clouds are masses of water droplets or ice crystals located in the troposphere layer of the atmosphere. Condensation is the basis by which clouds are formed, both in the sky and near the ground as fog. Clouds form when rising air cools to its **dew point**. The temperature at which condensation occurs is known as the dew point. In order for clouds to form, three ingredients are necessary: water vapor, condensation nuclei, and cooling. As water vapor cools, it reaches saturation point and condenses around **condensation nuclei**, tiny particles of dust, smoke, and other particulates within the atmosphere. Condensed water appears as tiny **cloud droplets**. Large groups of tiny water drops appear as clouds. All clouds are formed by the same physical process; however, the types of clouds will vary depending on atmospheric conditions. Depending upon the temperature, the tiny water droplets may freeze or remain in a liquid state. That is why most high-level clouds are tiny particles of ice. Closer to the earth, water vapor condenses on a surface, such as grass, and can be observed in the form of **dew** or frost, if the temperature is below freezing.

Humidity is moisture in the air. **Relative humidity** (reported as a percentage) is defined as the amount of moisture in the air relative to what it could hold if it were entirely saturated at a given temperature. Air saturated with water vapor has a relative humidity of 100%. In order for clouds to form, the moisture (humidity) in the air must condense. When humidity is high, rain clouds or fog are more likely to form. Air at different temperatures is capable of holding varying amounts of moisture. For example, as air cools, it is not able to hold as much moisture. Relative humidity tends to rise at night and decrease during the day, despite the amount of moisture remaining the same; the temperature is what varies. Relative humidity is important to know when watching for precipitation.

Cloud types vary depending on atmospheric conditions. They are classified by their shape and height. Stratus, cumulus, and cirrus are the three main cloud types.

Stratus clouds form at low altitudes. They form layers of sheet-like clouds and are associated with steady rains, snow, and very cloudy days. Stratus clouds at ground level are called fog.

Cumulus clouds usually form at mid-altitudes. These flat-bottomed, fluffy clouds are associated with fair weather.

Cirrus clouds form at high altitudes; they are long and feathery and are made of ice crystals. They are often seen before a weather change, such as rain or snow.

Clouds sometimes form in odd mixtures that look like combinations of the basic three types. The illustration below shows some of the "hybrid" clouds.

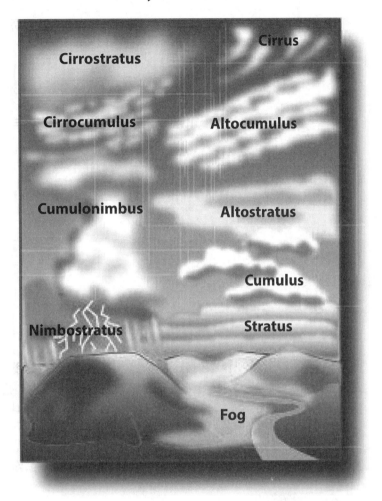

Sometimes we see lines in the sky left by airplanes. Condensing water vapor from the exhaust of jet engines forms condensation trails, often called **contrails**. These contrails can tell us a lot about conditions in the atmosphere. If a contrail is very short and evaporates quickly, atmospheric moisture levels at that particular altitude must be relatively dry. If the contrail remains visible and spreads out across the sky, one can infer that atmospheric moisture levels at that particular altitude must be relatively high. The direction of winds aloft may also be inferred by noting the direction in which the contrail spreads from the original path of the plane. Cloud heights may also be inferred by gauging the vertical position of contrails as they intersect clouds.

Name: _____ Date: _____

Quick Check

Matching

_____ 1. dew point

_____ 2. contrails

_____ 3. relative humidity

_____ 4. fog

_____ 5. humidity

a. amount of moisture in the air relative to what it could hold if it were entirely saturated at a given temperature

b. stratus clouds at ground level

c. moisture in the air

d. temperature at which condensation occurs

e. condensing water vapor from the exhaust of jet engines

Fill in the Blanks

6. Cloud types vary depending on _____ conditions.

7. Air saturated with water vapor has a relative humidity of _____%.

8. When _____ is high, rain clouds or fog are more likely to form.

9. _____ _____ tends to rise at night and decrease during the day, despite the amount of moisture remaining the same.

10. _____, _____, and _____ are the three main cloud types.

Multiple Choice

11. These clouds usually form at mid-altitudes.
    a. stratus
    c. cirrus
    b. cumulus
    d. cirrocumulus

12. These clouds are long and feathery.
    a. cumulonimbus
    c. cumulus
    b. cirrus
    d. stratus

13. Which of the following is NOT an ingredient in cloud formation?
    a. water vapor
    c. contrails
    b. condensation nuclei
    d. cooling

14. These fluffy clouds are associated with fair weather.
    a. cirrus
    c. stratus
    b. cumulus
    d. cirrocumulus

Name: _____ Date: _____

Knowledge Builder

Activity #1: Dew Point

Directions: Measure and record the air temperature (°C) in the data table below. Fill a soup can approximately half-full of water. Measure and record the temperature of the water. Gradually add ice to the water, stirring continuously. Warning: Do not stir with the thermometer. Measure and record the temperature of the ice water every 30 seconds for four minutes. Keep adding ice as needed.

Temperature (°C)									
Air	Starting Ice water	30 sec.	1 min.	1.5 min.	2 min.	2.5 min.	3 min.	3.5 min.	4 min.

Observation: At what time and temperature did condensation appear on the outside of the can?

_____

Conclusion: _____

_____

_____

_____

Activity #2: Cloud in a Bottle

Directions: Caution: this activity needs close teacher supervision. Place one to two ounces of water in a plastic 2-liter soda bottle. Swish the water completely around in the bottle. Pour most of the water out of the bottle, leaving only enough to cover the bottom of the bottle. Light a match; after it burns for a second or two, hold the match near the opening of the bottle, and blow the match out. Catch some of the smoke of the match in the bottle. Quickly put the cap tightly on the bottle. Squeeze the side of the bottle hard. Quickly release the pressure on the bottle. Hold the bottle against a dark background and look for a thin cloud to form in the bottle. Continue to squeeze and release the bottle a number of times in order to observe the cloud being formed. Check the cloud 20 minutes later.

Observation: Describe what you observed. _____

_____

_____

Conclusion: How is your cloud model similar to the way clouds form in nature? _____

_____

_____

Name: _____   Date: _____

Knowledge Builder

Activity #3: Cloud Observation

Directions: Each day, observe and record the characteristics of the clouds in the sky. Record the temperature, relative humidity, barometric pressure, and the wind speed and direction. Using a cloud chart, identify the types of clouds you observed. Arrange your data in the form of a table. This will enable you to look for patterns in the data you have collected.

Observation: Did you observe a pattern in the data you collected? Explain. _____

_____

_____

Conclusion: What can you infer about cloud types and the kind of weather that occurs? _____

_____

_____

Activity #4: Make a Psychrometer

Directions: One way to measure relative humidity is by using a device called a psychrometer. Make a psychrometer and measure the relative humidity of different areas in your school such as a classroom, locker room, gymnasium, outside, cafeteria, or kitchen. Lay two thermometers side by side. Compare the temperatures to insure that they are properly calibrated. Cut a 2 cm section of hollow shoestring (a flat shoe string will not work) and slip it over the end of one of the thermometers. Soak the small piece of shoestring with water. Carefully, yet vigorously, fan the ends of the two thermometers with a manila folder for 3 minutes. Measure and record the temperatures (°C) on both thermometers. Calculate the "Depression Temperature" by subtracting the lowest temperature from the highest temperature of the dry and wet bulbs. Figure the relative humidity of the area by using the chart on page 39.

Location	Dry Bulb Temperature (°C)	Wet Bulb Temperature (°C)	Depression Temperature (°C)	Relative Humidity

Conclusion _____

_____

_____

Name: _____　　Date: _____

Relative Humidity Chart

Dry-Bulb Temp. (°C)	Relative Humidity (%) Difference Between Wet- and Dry-Bulb Temperature (°C)																			
	1°	2°	3°	4°	5°	6°	7°	8°	9°	10°	11°	12°	13°	14°	15°	16°	17°	18°	19°	20°
-10	67	35																		
-9	69	39	9																	
-8	71	43	15																	
-7	73	48	20																	
-6	74	49	25																	
-5	76	52	29	7																
-4	77	55	33	12																
-3	78	57	37	17																
-2	79	60	40	22																
-1	81	62	43	26	8															
0	81	64	46	29	13															
1	83	66	49	33	17															
2	84	68	52	37	22	7														
3	84	70	55	40	26	12														
4	85	71	57	43	29	16														
5	86	72	58	45	33	20	7													
6	86	73	60	48	35	24	11													
7	87	74	62	50	38	26	15													
8	87	75	63	51	40	29	19	8												
9	88	76	64	53	42	32	22	12												
10	88	77	66	55	44	34	24	15	6											
11	89	78	67	56	46	36	27	18	9											
12	89	78	68	58	48	39	29	21	12											
13	89	79	69	59	50	41	32	23	15	7										
14	90	79	70	60	51	42	34	26	18	10										
15	90	80	71	61	53	44	36	27	20	13	6									
16	90	81	71	63	54	46	38	30	23	15	8									
17	90	81	72	64	55	47	40	32	25	18	11									
18	91	82	73	65	57	49	41	34	27	20	14	7								
19	91	82	74	65	58	50	43	36	29	22	16	10								
20	91	83	74	66	59	51	44	37	31	24	18	12	6							
21	91	83	75	67	60	53	46	39	32	26	20	14	9							
22	92	83	76	68	61	54	47	40	34	28	22	17	11	6						
23	92	84	76	69	62	55	48	42	36	30	24	19	13	8						
24	92	84	77	69	62	56	49	43	37	31	26	20	15	10	5					
25	92	84	77	70	63	57	50	44	39	33	28	22	17	12	8					
26	92	85	78	71	64	58	51	46	40	34	29	24	19	14	10	5				
27	92	85	78	71	65	58	52	47	41	36	31	26	21	16	12	7				
28	93	85	78	72	65	59	53	48	42	37	32	27	22	18	13	9	5			
29	93	86	79	72	66	60	54	49	43	38	33	28	24	19	15	11	7			
30	93	86	79	73	67	61	55	50	44	39	35	30	25	21	17	13	9	5		
31	93	86	80	73	67	61	56	51	45	40	36	31	27	22	18	14	11	7		
32	93	86	80	74	68	62	57	51	46	41	37	32	28	24	20	16	12	9	5	
33	93	87	80	74	68	63	57	52	47	42	38	33	29	25	21	17	14	10	7	
34	93	87	81	75	69	63	58	53	48	43	39	35	30	28	23	19	15	12	8	5
35	94	87	81	75	69	64	59	54	49	44	40	36	32	28	24	20	17	13	10	7

Unit 8: Severe Weather
Teacher Information

Topic: Severe weather is weather that can potentially cause harm to people and the environment.

Concepts:
- Hurricanes are large, whirling low-pressure systems that form over warm ocean water in the tropics.
- Tornadoes are violently rotating columns of air, in contact with the surface of the earth, visible as a rope-shaped or wedge-shaped cloud.

Naïve Concepts:
- There are no hurricanes in the Pacific.
- A tornado is a violently rotating column of air, in contact with the ground, always visible as a funnel cloud.

Science Process Skills:
Students will **develop vocabulary** relating to severe weather findings. Students will **infer** that severe weather is weather that can potentially cause harm to people and the environment.

Lesson Planner:
1. <u>Directed Reading</u>: Introduce the concepts and essential vocabulary relating to hurricanes and tornadoes using the directed reading exercise on the Student Information pages.
2. <u>Assessment</u>: Evaluate student comprehension of the information in the directed reading exercise using the quiz located on the Quick Check page.
3. <u>Concept Reinforcement</u>: Strengthen student understanding of concepts with the activities found on the Knowledge Builder page. **Materials Needed:** Activity #2—two clear plastic 2-liter soda bottles, tape, water, food coloring

Extension: Each year hundreds of tornadoes touch down in the United States. Students research tornadoes and create a list of home safety tips and procedures.

Real World Application: When a tornado watch is issued, it means that a tornado is possible. A tornado warning is issued when a tornado has actually been spotted or is strongly indicated on radar.

Unit 8: Severe Weather
Student Information

Severe weather is weather that can potentially cause harm to people and the environment. Hurricanes and tornadoes are two examples of potentially dangerous and destructive storms.

Hurricanes are large, whirling low pressure systems that form over warm ocean water in the tropics. These storms often bring extremely high winds, thunder, lightning, large amounts of rain, and tornadoes. When hurricanes strike land, they may be very destructive. If north of the equator, they spin in a counterclockwise direction. The direction of spin is due to the earth's spin, called the Coriolis Effect.

Hurricanes usually begin in the Atlantic Ocean near the west coast of Africa as tropical cyclones. Here the ocean water is very warm. When a large mass of unstable warm air beings to rise

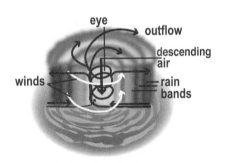

and cool, clouds begin to form. The rising mass of air starts to spin. If the spinning winds increase in speed, the unstable air mass becomes a tropical cyclone. As the cyclone is carried west by the northeast trade winds, it may develop into a larger storm. When winds speeding around the eye (area of clear skies in the center of the storm) reach 74 miles per hour, the storm is classified as a hurricane. Tropical cyclones whose winds are 74+ mph in the eastern Pacific are called "Eastern Pacific Hurricanes." In the western half of the Pacific, the most common term is "typhoon."

Wind speeds are used to categorize hurricanes. A wind speed scale is used for this purpose. Using the scale, the huge storm may be Category I (74–95 mph, 119–153 kph winds), Category II (96–110 mph, 154–177 kph winds) Category III (111–130 mph, 178–209 kph winds), Category IV (130–155 mph, 210–249 kph winds), and Category V (most severe storm, 155 mph+, 249+ kph).

Tornadoes are very destructive storms that form over land. They cause great damage with winds that spin up to 300 mph (480 kph). Although they may happen throughout the year, the spring and summer seasons are the most active times for tornadoes. The term "Tornado Alley "is used to refer to those parts of the Great Plains and Midwest where sightings are frequent.

Tornadoes occur in very severe thunderstorms where winds blow at different heights and at different speeds. This causes a funnel cloud to form. Tornado winds spiral, or spin, in a counterclockwise direction around the center or eye of a low pressure area. To be a tornado, the spinning cloud must extend from the cloud to the surface of the earth. The spinning cloud may be rope-shaped or wedge-shaped. Both have counterclockwise winds moving in and up around a low-pressure area that is the center, or eye.

Name: _____   Date: _____

Quick Check

Matching

_____ 1.   the eye

_____ 2.   tornadoes

_____ 3.   Category V

_____ 4.   wind speeds

_____ 5.   hurricanes

a.   used for categorizing hurricanes

b.   most severe hurricane

c.   large, whirling low-pressure systems that form over warm ocean water in the tropics

d.   area of clear skies in the center of the storm

e.   very destructive storms that form over land

Fill in the Blanks

6.   Hurricanes usually begin near the west coast of Africa as _____

_____.

7.   _____ _____ are used to categorize hurricanes.

8.   The term _____ _____ is used to refer to those parts of the Great Plains and Midwest where tornado sightings are frequent.

9.   _____ and _____ are two examples of potentially dangerous and destructive storms.

10. _____ _____ is weather that can potentially cause harm to people and the environment.

Multiple Choice

11. If a hurricane is north of the equator, what direction does it spin?

   a.   counterclockwise

   b.   clockwise

   c.   easterly

   d.   westerly

12. If a hurricane is classified as a Category I, what is the wind speed?

   a.   111–130 mph

   b.   74–95 mph

   c.   155+ mph

   d.   96–110 mph

13. A hurricane with wind speeds of 155+ mph is classified as a _____.

   a.   Category I

   b.   Category III

   c.   Category V

   d.   Category IV

14. In what seasons are tornadoes most likely to occur?

   a.   spring and summer

   b.   fall and winter

   c.   summer and fall

   d.   spring and fall

Name: _____ Date: _____

Knowledge Builder

Activity #1: Hurricanes and Tornadoes

Directions: Each year tornadoes and hurricanes cause thousands of dollars of damage. Explain the difference in these two destructive storms. Complete the data table below.

Storm	How the Storm Forms	Description	Types of damage
1. tornado			
2. hurricane			

Activity #2: Tornado in a Bottle

Directions: Fill one 2-liter plastic soda bottle three-quarters full of water. Add a couple drops of food coloring. Invert the other bottle so that it sits on top of the first bottle with the openings together. Wind the tape tightly around the necks of the bottles so that no water can leak out. The top bottle should be securely balanced on top of the bottom bottle. Now, hold the bottles with two hands and swirl the water around. Turn the bottles upside down and observe what happens.

Observation: _____

_____

_____

_____

_____

_____

Unit 9: Mapping Weather
Teacher Information

Topic: Symbols are used to represent weather conditions on a map.

Concepts:
- To make weather maps and forecast weather, weather data must be collected.
- Symbols are used to represent weather conditions on a map.

Naïve Concepts:
- Meteorologists use only computers to generate weather maps.
- Forecasting the weather is a subject that is learned and mastered perfectly.

Science Process Skills:

    Students will gather and plot real-world weather data both from where they live and from many cities in the United States. Students will be **communicating** and **developing vocabulary**. Students will be **creating models** of weather maps during the process of **analyzing** and **interpreting** data.

Lesson Planner:

1. <u>Directed Reading</u>: Introduce the concepts and essential vocabulary relating to mapping weather using the directed reading exercise found on the Student Information page.

2. <u>Assessment</u>: Evaluate student comprehension of the information in the directed reading exercise using the quiz located on the Quick Check page.

3. <u>Concept Reinforcement</u>: Strengthen student understanding of concepts with the activities found on the Knowledge Builder page.

Extension: Various weather factors (extreme cold or heat, drought, excess rainfall, etc.) are likely to impact the economy indirectly. Students consider one of the factors and investigate its effect on the local economy.

Real World Application: The floods of 1927 and 1993 on the Mississippi River are considered the most destructive river floods in the history of the United States.

Unit 9: Mapping Weather
Student Information

Weather data must be collected from the earth's surface and atmosphere in order to make weather maps and forecast weather. After weather data has been collected, meteorologists are able to plot and interpret this information to make a skilled weather forecast.

Weather data is collected in various ways. **Weather satellites**, or meteorological satellites, orbit high above Earth's surface in the atmosphere. They make images of clouds and storms and can track their movement. **Radar** (**ra**dio **d**etecting **a**nd **r**anging) instruments send out radio signals that are reflected from objects such as clouds and rain. The reflections are then turned into images. Ground level radar identifies areas of heavy, medium, or low precipitation. On television, radar images are different colors to show different amounts of precipitation.

Meteorologists truly have a language of their own. **Weather boxes** are simply an organized collection of symbols used to represent the data. Weather boxes consist of a variety of weather data and are used to represent the weather conditions for a specific data collection location. This data is collected at National Weather Service stations via weather balloons, which are launched roughly every eight hours. Weather boxes are positioned near the city on a weather map and contain the following information:

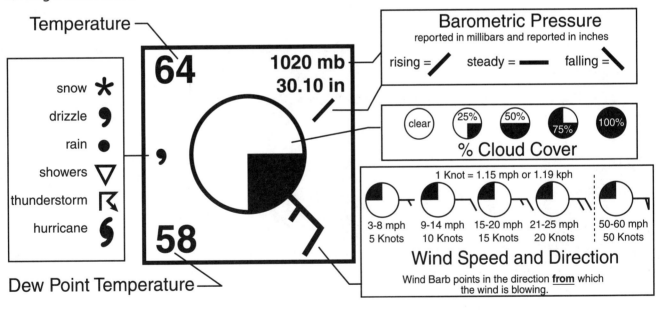

More sophisticated weather boxes may include cloud types, visibility, and the amount of change in barometric pressure in the past three hours. Most weather symbols have variations of the symbol to represent the intensity of the data.

Name: _____ Date: _____

Quick Check

Matching

_____ 1.   weather satellites

_____ 2.   weather box

_____ 3.   radar

_____ 4.   weather data

_____ 5.   ground level radar

a.   identifies areas of heavy, medium, or low precipitation

b.   radio detecting and ranging

c.   meteorological satellites

d.   an organized collection of symbols used to represent data

e.   collected at National Weather Service stations via weather balloons

Fill in the Blanks

6.   _____ _____ must be collected from the earth's surface and atmosphere in order to make weather maps and forecast weather.

7.   Most _____ _____ have variations of the symbol to represent the intensity of the data.

8.   _____ _____ are simply an organized collection of symbols used to represent the data.

9.   On television, _____ _____ are different colors to show different amounts of precipitation.

10. Weather satellites, or meteorological satellites, are orbiting high above Earth's surface in the _____.

Multiple Choice

11. This weather symbol represents steady barometric pressure.

a.   b.   c.   d.   ——

12. This weather symbol represents cloud cover.

a.   b.   c.   d.

13. This weather symbol represents wind speed and direction.

a.   b.   c.   d.

Name: _____ Date: _____

Knowledge Builder

Activity: Charting Weather Data

Directions: Use the information provided in the weather boxes on the map on page 48 to answer the following questions.

1. What is the temperature in Des Moines, Iowa? _____

2. What is the wind speed in Atlanta, Georgia? _____

3. What type of a weather warning is likely being issued in Miami, Florida? _____
    _____

4. What is the wind direction in Oklahoma City, Oklahoma? _____

5. What it the barometric pressure in Columbus, Ohio? _____

6. Is the barometric pressure rising, falling, or remaining steady in Kansas City, Missouri? How do you know? _____
    _____

7. What is the percentage of cloud cover in Memphis, Tennessee? _____

8. What type of precipitation is occurring in St. Louis, Missouri? _____

9. From your understanding of barometric pressure, identify and label an area of stormy weather with very low pressure on the map.

10. Based on your understanding of the weather conditions in Miami, Florida, what type of weather is likely to occur in Tampa in several hours? Explain your reasoning. _____
    _____
    _____

11. Plot the data (from the chart below) on the map for the cities that are missing weather boxes. This time, you will need to create the weather boxes.

	City	Cloud Cover	Temperature	Barometric Pressure	Type of Weather	Wind Speed/ Directions
a.	San Antonio, TX	Clear	80	30.0 ⏱	Sunny	5 mph SW
b.	Denver, Co	100%	45	29.1 ▯	Snow	15 mph W
c.	Milwaukee, WI	50%	35	30.1 steady		11 mph NW
d.	Tampa, FL	75%	83	29.7 ▯	Thunderstorm	20 mph SE

Name: _____ Date: _____

Weather Map

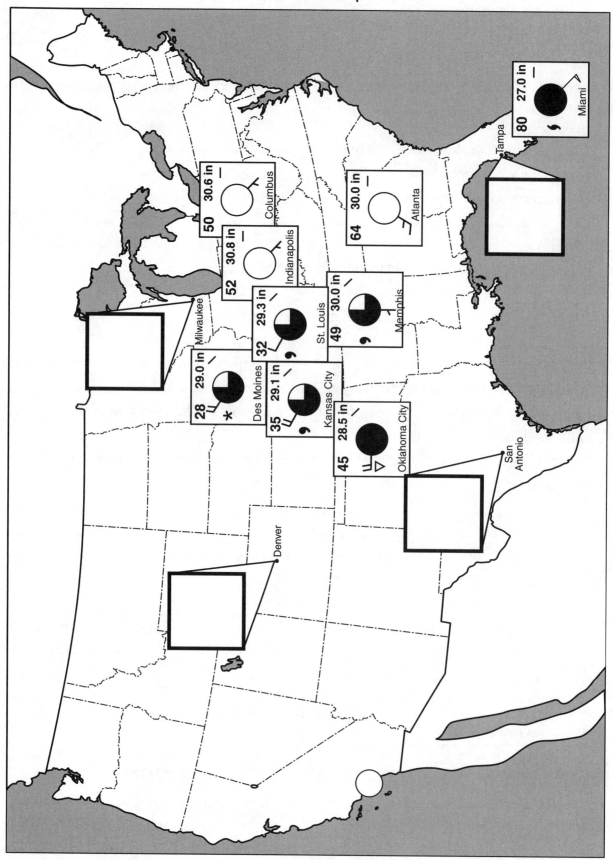

Unit 10: Weather Fronts
Teacher Information

Topic: Global patterns of atmospheric movement influence local weather.

Standards:
    NSES Unifying Concepts and Processes, (D)
    NCTM Geometry, Measurement, and Data Analysis and Probability
    See **National Standards** section (pages 61–65) for more information on each standard.

Concepts:
- Fronts are where active weather occurs.
- Temperature is dependent upon the air mass present.

Naïve Concepts:
- Temperature will always be warmer in the afternoon than in the morning.
- The same weather is occurring everywhere at the same time.

Science Process Skills:
    Students will make **observations** about air masses and the type of weather associated with the different front formations. Students will be **communicating** and **developing vocabulary** during the process of **collecting**, **recording**, **analyzing**, and **interpreting** data. They will **infer** weather conditions are related to movement of cold and warm air masses.

Lesson Planner:
1. <u>Directed Reading</u>: Introduce the concepts and essential vocabulary relating to weather fronts using the directed reading exercise found on the Student Information page.
2. <u>Assessment</u>: Evaluate student comprehension of the information in the directed reading exercise using the quiz located on the Quick Check page.
3. <u>Concept Reinforcement</u>: Strengthen student understanding of concepts with the activities found on the Knowledge Builder page. **Materials Needed:** political map of United States, colored pencils

Extension: Provide students with weather poems by various authors such as Carl Sandburg's "The Fog," Myra Cohn Livingston's "Coming Storm," or Christina G. Rossetti's "Who Has Seen the Wind?" Instruct students to create their own weather poems.

Real World Application: On September 13, 2008, at 2:10 A.M., a Category II hurricane hit Galveston, Texas. Hurricane Ike, with winds up to 110 miles per hour, was the third most destructive hurricane to ever make landfall in the United States.

Unit 10: Weather Fronts
Student Information

When the sun pumps lots of heat into Earth's atmosphere, it creates air masses with many different temperatures and humidities. As these air masses bump and jostle each other, they create weather. Meteorologists define air masses by where they form. **Maritime air masses** assemble over oceans. **Continental air masses** build over land.

North America contends with four types of air masses:

1. **Maritime tropical** air masses form over the ocean near the equator and can bring hot, humid summers or stormy winters if they bang into cold, northerly air heading south.
2. **Maritime polar** air masses form over the Pacific Ocean and the North Atlantic. They carry cool, moist air.
3. **Continental tropical** air masses occur in the summer and form over Mexico. They bring hot air to the southwestern states.
4. **Continental polar** air masses form over northern Canada and may cause the mercury in thermometers in northern states to nose-dive.

Points of contact between air masses are called **fronts**. Fronts are found along leading edges of air masses. The temperature and pressure of the advancing air mass dictates the name of the front. For example, if a cold air mass overtakes a warm air mass, a **cold front** forms. Typically, cold, dense air plows under warm, moist air, causing it to rise rapidly. Such a front is likely to yield a sudden, heavy rain shower as the warm, moist air is quickly cooled when it rises. Conversely, if a warm air mass overtakes a cold air mass, a **warm front** is created. The less dense, warm air tends to slide over the heavy, dense cooled air. Stratus clouds often occur with longer periods of steady rainfall. A **stationary front** occurs when air masses tend to remain in place for a period of time. When two cold air masses collide and push a warm air mass up between them, an **occluded front** forms. Weather is hard to predict in this type of front. Identifying and tracking fronts is an important skill in accurately predicting the weather.

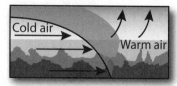

A cold front usually moves from northwest to southeast.

A warm front usually moves from southwest to northeast.

Fronts are where active weather occurs. Most weather systems move from either the west-southwest to east-northeast or from west-northwest to east-southeast. At times, weather patterns will be stable and yield little or no significant changes. If the differences in the air masses are great, storms are spawned. Strong cold fronts moving quickly into warm, moist air often result in strong storms and yield a variety of severe weather conditions.

- **Rainstorms** and **snowstorms** result from the collision of different fronts. Nimbostratus clouds form when a warm front moves in and rises over cold air, often resulting in heavy rain or snow. A **blizzard** results when wind speeds exceed 56 km/h and the temperature is less than -7°C.

- **Thunderstorms** form when a cold front moves in and meets a warm front. High, cumulonimbus clouds produce thunder, lightning, and sudden air movements called **wind shears,** which are dangerous for planes.

- Spinning air masses called **cyclones** form when cool air swoops in to replace rising warm air in a region of low pressure. **Anticyclones** form in high-pressure areas with cold, dry air that spirals out in a direction opposite to cyclones, usually bringing clear, dry, and fair weather.

- Cyclones called **hurricanes** form powerful storms over tropical oceans. Similar storms that form over the western Pacific Ocean are called **typhoons**. The centers, or eyes, of such storms are calm, while around them air hurtles by at up to 480 km/h.

- Whirling funnel clouds over land called **tornadoes** form in low cumulonimbus clouds. **Water spouts** form over water. Both types of storms may cause great damage with winds that spin up to 480 km/h for tornadoes and 95 km/h for water spouts.

One way meteorologists identify air masses is to chart relative temperatures. Points with the same temperature are connected with a line on the weather map. These lines of equal temperatures are called **isotherms**. After their weather data has been collected, meteorologists are able to plot and interpret this information to make a skilled weather forecast. Meteorologists use symbols to represent fronts on a weather map. The direction that the spurs (points) or bubbles are pointing on the front indicates the direction in which the front is moving.

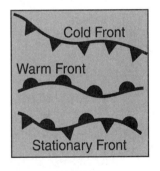

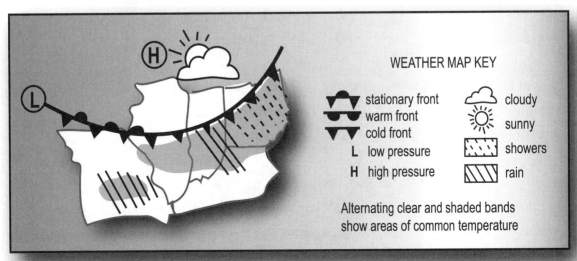

Name: _____   Date: _____

Quick Check

Matching

_____ 1. fronts   a. hurricanes

_____ 2. continental air masses   b. usually move from northwest to southeast

_____ 3. cyclones   c. points of contact between air masses

_____ 4. cold fronts   d. assemble over oceans

_____ 5. maritime air masses   e. build over land

Fill in the Blanks

6. A _____ _____ occurs when air masses tend to remain in place for a period of time.

7. _____ _____ air masses form over the Pacific Ocean and the North Atlantic.

8. If a cold air mass overtakes a warm air mass, a _____ _____ forms.

9. Whirling funnel clouds over land called _____ form in low cumulonimbus clouds.

10. _____ _____ air masses form over the ocean near the equator and can bring hot, humid summers or stormy winters if they bang into cold, northerly air heading south.

Multiple Choice

11. Hurricanes that form over the western Pacific Ocean are called _____.

    a. wind shears   b. tornadoes

    c. typhoons   d. water spouts

12. This type of air mass forms over northern Canada.

    a. continental polar   b. continental tropical

    c. maritime polar   d. maritime tropical

13. This storm results when wind speeds exceed 56 km/h and the temperature is less than -7°C.

    a. typhoon   b. cyclone

    c. blizzard   d. hurricane

Name: _____   Date: _____

Knowledge Builder

Activity: Warm and Cold Air Masses

Before beginning the activity, take some time to review some important weather information you learned in previous lessons.
- **Review the information contained in the weather box found on page 45.**
- **Review the symbols for cold fronts and warm fronts. Remember the spurs or bubbles indicate the direction in which the front is moving.**

Directions: Examine the temperature data on the chart provided below. Draw a weather box around each city on the weather map on page 54, and transfer the data below to its appropriate location on the map. Next, draw contour lines on the map that connect points of equal temperature. These lines are called isotherms. Track the isotherms from various cities: two distinct air masses should emerge. Label the relative temperature of each air mass warm or cold. Use colored pencils to represent where you believe a front exists. From what you have learned in previous lessons about pressure or air masses, label one air mass "low" and one air mass "high."

City	Temperature	City	Temperature
Chicago, IL	65	Memphis, TN	85
Cincinnati, OH	83	Minneapolis, MN	60
Des Moines, IA	65	Nashville, TN	80
Detroit, MI	79	Oklahoma City, OK	69
Little Rock, AR	68	Springfield, IL	65
Madison, WI	64	St. Louis, MO	67

1. Determine whether the cold air mass is moving into the area of warm air or vice versa. Explain how you know. _____

   _____

   _____

   _____

2. Along the border of the air masses on the map draw a symbol for either the cold front or warm front.

3. Predict what type of weather is likely to occur along the front you have just drawn.

   _____

   _____

   _____

Name: _____   Date: _____

Weather Map

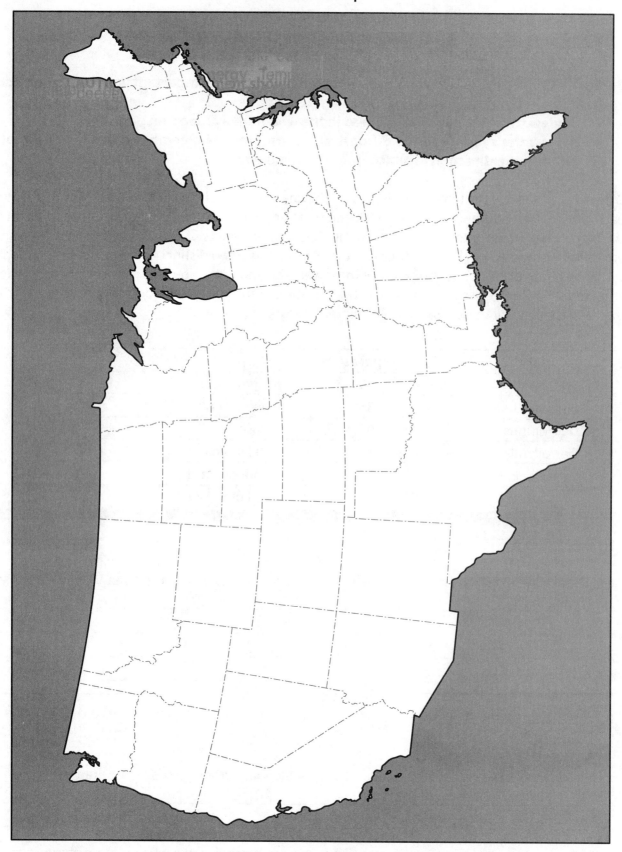

Unit 11: Climate
Teacher Information

Topic: Several factors affect the climate of an area.

Standards:
   NSES Unifying Concepts and Processes, (D)
   NCTM Number and Operations and Measurement
   See **National Standards** section (pages 61–65) for more information on each standard.

Concepts:
- Climate is the general weather of an area over a long period of time.

Naïve Concepts:
- Weather and climate are the same thing.

Science Process Skills:
   Students will **develop vocabulary** relating to climate. Students will **infer** several factors affect the climate of an area.

Lesson Planner:

1. Directed Reading: Introduce the concepts and essential vocabulary relating to climate using the directed reading exercise found on the Student Information page.

2. Assessment: Evaluate student comprehension of the information in the directed reading exercise using the quiz located on the Quick Check page.

3. Concept Reinforcement: Strengthen student understanding of concepts with the activities found on the Knowledge Builder page. **Materials Needed:** copy paper, scissors, and glue

Extension: Deforestation is the logging and/or burning of a large number of trees in an area so the land can be used for something instead of a forest. Students research deforestation to determine how it can affect climate.

Real World Application: El Niño is a climate condition that occurs every few years. In an El Niño year, weather patterns are changed because of an abnormal warming of the Pacific Ocean. El Niño is associated with flooding and droughts and other weather disturbances around the world.

Climate
Student Information

The temperatures and precipitation for a particular area over a long period of time is called the area's **climate**. Climate includes seasonal changes in weather. Both temperature and precipitation are affected by several factors that lead to a rich variety of climates in our "water world" planet.

Temperature varies largely because of three things.

1. **Latitude**, or distance of a place north or south of the equator, affects temperature. Temperatures are generally lower as you get farther from the equator (higher latitudes).

2. **Elevation**, or altitude, of an area is its distance above sea level. Air thins as you climb a mountain; thin air holds less heat. Therefore, temperatures usually decrease as elevation increases.

3. **Temperature of ocean currents** directly affects the temperature of the air above them. In general, warm ocean currents flow away from the equator, and cool currents flow toward the equator. Major currents like the Gulf Stream (see map) can significantly warm the air near landmasses that would otherwise be quite cold.

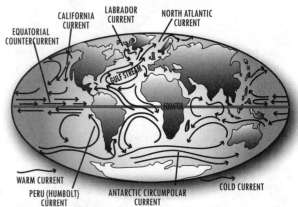

Precipitation varies because of two major factors.

1. **Prevailing winds** are winds that blow more often from one direction than the other. They may be warm or cold and carry varying amounts of water, depending on whether they are blowing off water or land. Sometimes deserts can even exist near large bodies of water if the prevailing winds are off a large landmass; such is the case with the Sahara Desert in northern Africa.

2. **Mountain ranges** alter the flow of prevailing winds. Air must rise to get over a mountain, and rising air cools and becomes incapable of holding as much water. Therefore, the side of a mountain facing the prevailing wind (the windward side) tends to get a lot of moisture. The leeward side of the same mountain gets sinking air stripped of most of its moisture, often resulting in a desert in the mountain's so-called "rain shadow." Such is the case for the Sierra Nevadas of western Northern America and the desert-dry Great Basin east of it.

The earth can be divided into three large climate zones based on average annual temperatures and defined by latitude.

1. **Polar**, or **arctic, climate zones** exist between latitudes of 90° and 60°. Here, near either pole, there is no summer and the average yearly temperature remains below 0°C. If you travel to the icecaps of Greenland or the continent of Antarctica, you will experience a polar zone.

2. **Temperate zones** extend from latitudes 60° to 30°. Here there are seasonal swings in temperature, and the average temperatures can vary from 5° to 20°C. Expect to find inland deserts that can be quite hot and dry during the day, but cool, or even cold, at night.

3. The **tropical zone** bands the earth from the equator at 0° to 30° on either side of the equator. Temperatures are hot, there is high humidity, and there are no winters. Even during the coldest months, temperatures never drop below 18°C. Because the trade winds blow east to west, you will find deserts on the west sides of north-south trending mountain ranges. The Atacama Desert of Chile and Peru is exceedingly dry and cold because of cool offshore ocean currents.

Within each zone, **marine** and **continental climates** provide variation. The large bodies of water around marine climates give up moisture to winds, providing more precipitation. Summers tend to be warm (rather than hot) and winters mild. Dryer continental climates exist within large landmasses. Temperatures swing more wildly from high to low. Summers may be hot, and winters may be quite cold.

While we owe our climate zones to Earth's round shape, the annual variation in temperature and precipitation we call the four seasons—spring, summer, fall, and winter—results from the fact that our planet is tilted at an angle of 23.5° relative to the sun. When the Northern Hemisphere of the earth is tilted toward the sun, that hemisphere gets more heat and experiences summer, while the Southern Hemisphere deals with winter. Six months later, the situation is reversed. In the months between summer and winter, both hemispheres get the same amount of sunlight and experience either spring or fall, depending on whether the transition is toward summer or winter. **Spring** in the Northern Hemisphere begins March 20 or 21, and **autumn** begins September 22 or 23.

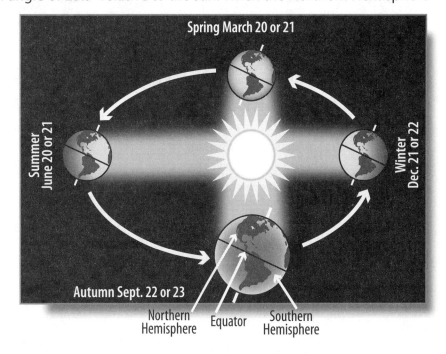

Name: _____   Date: _____

Quick Check

Matching

_____ 1.   climate

_____ 2.   elevation

_____ 3.   temperate zones

_____ 4.   latitude

_____ 5.   spring

a.   begins March 20 or 21

b.   extend from latitudes 60° to 30°

c.   altitude

d.   temperatures and precipitation for a particular area over a long period of time

e.   distance of a place north or south of the equator

Fill in the Blanks

6.   _____ _____ alter the flow of prevailing winds.

7.   Temperature of _____ _____ directly affects the temperature of the air.

8.   Within each zone, _____ and _____ _____ provide variation.

9.   _____ _____ are winds that blow more often from one direction than the other.

10. The earth can be divided into three large _____ _____ based on average annual temperatures and defined by latitude.

Multiple Choice

11. When does autumn begin in the Northern Hemisphere?
    a.   October 22 or 23   b.   February 20 or 21
    c.   March 20 or 21   d.   September 22 or 23

12. This climate zone bands the earth from the equator at 0° to 30° on either side of the equator.
    a.   temperate   b.   tropical
    c.   arctic   d.   marine

13. This climate zone exists between latitudes of 90° and 60°.
    a.   arctic   b.   marine
    c.   tropical   d.   temperate

Name: _____ Date: _____

Knowledge Builder

Activity: Climate Zone Tri-Fold Booklet

Directions: Fold an 8 1/2" X 11" piece of paper as shown. Fold each side to the midpoint and crease. Now center and glue the climate zone map over the opening of the booklet. Carefully cut through the map and opening. Open booklet and label each of the three sections: Polar Zone, Tropical Zone, and Temperate Zone. Describe each climate zone including the location, temperature, and seasons.

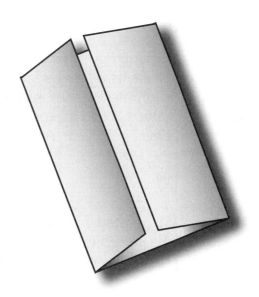

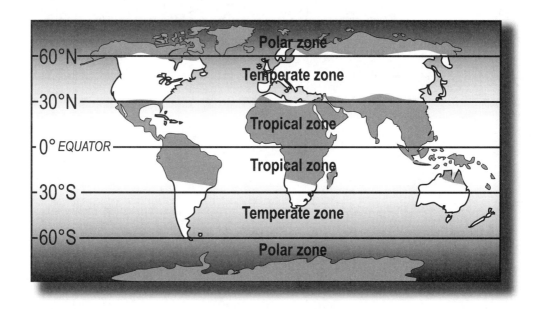

Name: _____   Date: _____

Inquiry Investigation Rubric

Category	4	3	2	1
Participation	Used time well, cooperative, shared responsibilities, and focused on the task.	Participated, stayed focused on task most of the time.	Participated, but did not appear very interested. Focus was lost on several occasions.	Participation was minimal OR student was unable to focus on the task.
Components of Investigation	All required elements of the investigation were correctly completed and turned in on time.	All required elements were completed and turned in on time.	One required element was missing/or not completed correctly.	The work was turned in late and/or several required elements were missing and/or completed incorrectly.
Procedure	Steps listed in the procedure were accurately followed.	Steps listed in the procedure were followed.	Steps in the procedure were followed with some difficulty.	Unable to follow the steps in the procedure without assistance.
Mechanics	Flawless spelling, punctuation, and capitalization.	Few errors.	Careless or distracting errors.	Many errors.

Comments:

National Standards in Science, Math, and Technology

NSES Content Standards (NRC, 1996)
National Research Council (1996). *National Science Education Standards.* Washington, D.C.: National Academy Press.

UNIFYING CONCEPTS: K–12
Systems, Order, and Organization: The natural and designed world is complex. Scientists and students learn to define small portions for the convenience of investigation. The units of investigation can be referred to as systems. A system is an organized group of related objects or components that form a whole. Systems can consist of machines.

Systems, Order, and Organization
The goal of this standard is to ...
- Think and analyze in terms of systems.
- Assume that the behavior of the universe is not capricious. Nature is predictable.
- Understand the regularities in a system.
- Understand that prediction is the use of knowledge to identify and explain observations.
- Understand that the behavior of matter, objects, organisms, or events has order and can be described statistically.

Evidence, Models, and Explanation
The goal of this standard is to ...
- Recognize that evidence consists of observations and data on which to base scientific explanations.
- Recognize that models have explanatory power.
- Recognize that scientific explanations incorporate existing scientific knowledge (laws, principles, theories, paradigms, models) and new evidence from observations, experiments, or models.
- Recognize that scientific explanations should reflect a rich scientific knowledge base, evidence of logic, higher levels of analysis, greater tolerance of criticism and uncertainty, and a clear demonstration of the relationship between logic, evidence, and current knowledge.

Change, Constancy, and Measurement
The goal of this standard is to ...
- Recognize that some properties of objects are characterized by constancy, including the speed of light, the charge of an electron, and the total mass plus energy of the universe.
- Recognize that changes might occur in the properties of materials, position of objects, motion, and form and function of systems.
- Recognize that changes in systems can be quantified.
- Recognize that measurement systems may be used to clarify observations.

National Standards in Science, Math, and Technology (cont.)

Form and Function

The goal of this standard is to …

- Recognize that the form of an object is frequently related to its use, operation, or function.
- Recognize that function frequently relies on form.
- Recognize that form and function apply to different levels of organization.
- Enable students to explain function by referring to form, and explain form by referring to function.

NSES Content Standard A: Inquiry

- Abilities necessary to do scientific inquiry
    - Identify questions that can be answered through scientific investigations.
    - Design and conduct a scientific investigation.
    - Use appropriate tools and techniques to gather, analyze, and interpret data.
    - Develop descriptions, explanations, predictions, and models using evidence.
    - Think critically and logically to make relationships between evidence and explanations.
    - Recognize and analyze alternative explanations and predictions.
    - Communicate scientific procedures and explanations.
    - Use mathematics in all aspects of scientific inquiry.
- Understanding about inquiry
    - Different kinds of questions suggest different kinds of scientific investigations.
    - Current scientific knowledge and understanding guide scientific investigations.
    - Mathematics is important in all aspects of scientific inquiry.
    - Technology used to gather data enhances accuracy and allows scientists to analyze and quantify results of investigations.
    - Scientific explanations emphasize evidence, have logically consistent arguments, and use scientific principles, models, and theories.
    - Science advances through legitimate skepticism.
    - Scientific investigations sometimes result in new ideas and phenomena for study, generate new methods or procedures, or develop new technologies to improve data collection.

NSES Content Standard B: Properties and Changes of Properties in Matter 5–8

NSES Content Standard D: Structure of the Earth System 5–8

NSES Content Standard D: Earth in the Solar System 5–8

NSES Content Standard E: Science and Technology 5–8

- Abilities of technological design
    * Identify appropriate problems for technological design.
    * Design a solution or product.
    * Implement the proposed design.
    * Evaluate completed technological designs or products.
    * Communicate the process of technological design.

National Standards in Science, Math, and Technology (cont.)

- Understanding about science and technology
    * Scientific inquiry and technological design have similarities and differences.
    * Many people in different cultures have made and continue to make contributions.
    * Science and technology are reciprocal.
    * Perfectly designed solutions do not exist.
    * Technological designs have constraints.
    * Technological solutions have intended benefits and unintended consequences.

NSES Content Standard F: Science in Personal and Social Perspectives 5–8

- Science and technology in society
    * Science influences society through its knowledge and world view.
    * Societal challenges often inspire questions for scientific research.
    * Technology influences society through its products and processes.
    * Scientists and engineers work in many different settings.
    * Science cannot answer all questions, and technology cannot solve all human problems.

NSES Content Standard G: History and Nature of Science 5–8

- Science as human endeavor
- Nature of science
    * Scientists formulate and test their explanations of nature using observation, experiments, and theoretical and mathematical models.
    * It is normal for scientists to differ with one another about interpretation of evidence and theory.
    * It is part of scientific inquiry for scientists to evaluate the results of other scientists' work.
- History of science
    * Many individuals have contributed to the traditions of science.
    * Science has been and is practiced by different individuals in different cultures.
    * Tracing the history of science can show how difficult it was for scientific innovators to break through the accepted ideas of their time to reach the conclusions we now accept.

National Standards in Science, Math, and Technology (cont.)

Standards for Technological Literacy (STL) ITEA, 2000
International Technology Education Association (2000). *Standards for Technological Literacy.* Reston, VA: International Technology Education Association.

The Nature of Technology
Students will develop an understanding of the:
1. Characteristics and scope of technology.
2. Core concepts of technology.
3. Relationships among technologies and the connections between technology and other fields of study.

Technology and Society
Students will develop an understanding of the:
4. Cultural, social, economic, and political effects of technology.
5. Effects of technology on the environment.
6. Role of society in the development and use of technology.
7. Influence of technology on history.

Design
Students will develop an understanding of the:
8. Attributes of design.
9. Engineering design.
10. Role of troubleshooting, research and development, invention and innovation, and experimentation in problem solving.

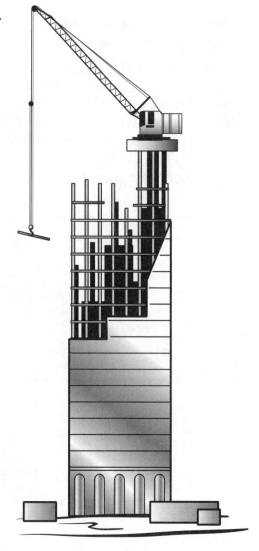

Abilities for a Technological World
Students will develop abilities to:
11. Apply the design process.
12. Use and maintain technological products and systems.
13. Assess the impact of products and systems.

The Designed World
Students will develop an understanding of and be able to select and use:
14. Medical technologies.
15. Agricultural and related biotechnologies.
16. Energy and power technologies.
17. Information and communication technologies.
18. Transportation technologies.
19. Manufacturing technologies.
20. Construction technologies.

National Standards in Science, Math, and Technology (cont.)

Principles and Standards for School Mathematics (NCTM), 2000

National Council for Teachers of Mathematics (2000). *Principles and Standards for School Mathematics.* Reston, VA: National Council for Teachers of Mathematics.

Number and Operations
Students will be enabled to:
- Understand numbers, ways of representing numbers, relationships among numbers, and number systems.
- Understand meanings of operations and how they relate to one another.
- Compute fluently and make reasonable estimates.

Algebra
Students will be enabled to:
- Understand patterns, relations, and functions.
- Represent and analyze mathematical situations and structures using algebraic symbols.
- Use mathematical models to represent and understand quantitative relationships.
- Analyze change in various contexts.

Geometry
Students will be enabled to:
- Analyze characteristics and properties of two- and three-dimensional geometric shapes and develop mathematical arguments about geometric relationships.
- Specify locations and describe spatial relationships using coordinate geometry and other representational systems.
- Apply transformations and use symmetry to analyze mathematical situations.
- Use visualization, spatial reasoning, and geometric modeling to solve problems.

Measurement
Students will be enabled to:
- Understand measurable attributes of objects and the units, systems, and processes of measurement.
- Apply appropriate techniques, tools, and formulas to determine measurements.

Data Analysis and Probability
Students will be enabled to:
- Formulate questions that can be addressed with data and collect, organize, and display relevant data to answer them.
- Select and use appropriate statistical methods to analyze data.
- Develop and evaluate inferences and predictions that are based on data.
- Understand and apply basic concepts of probability.

Science Process Skills

Introduction: Science is organized curiosity, and an important part of this organization includes the thinking skills or information-processing skills. We ask the question "why?" and then must plan a strategy for answering the question or questions. In the process of answering our questions, we make and carefully record observations, make predictions, identify and control variables, measure, make inferences, and communicate our findings. Additional skills may be called upon, depending on the nature of our questions. In this way, science is a verb, involving active manipulation of materials and careful thinking. Science is dependent on language, math, and reading skills, as well as the specialized thinking skills associated with identifying and solving problems.

BASIC PROCESS SKILLS:

Classifying: Grouping, ordering, arranging, or distributing objects, events, or information into categories based on properties or criteria, according to some method or system.

> *Example:* Classifying cloud types, snowflakes, forms of precipitation.

Observing: Using the senses (or extensions of the senses) to gather information about an object or event.

> *Example:* Observing wind direction, cloud formation, whether it is sunny or cloudy, whether it is raining or clear.

Measuring: Using both standard and nonstandard measures or estimates to describe the dimensions of an object or event. Making quantitative observations.

> *Example:* Measuring temperature, measuring the amount of precipitation, measuring wind speed and direction, measuring relative humidity.

Inferring: Making an interpretation or conclusion based on reasoning to explain an observation.

> *Example:* Stating that because barometric pressure is falling rapidly, a storm is coming.

Communicating: Communicating ideas through speaking or writing. Students may share the results of investigations, collaborate on solving problems, and gather and interpret data both orally and in writing. Using graphs, charts, and diagrams to describe data.

> *Example:* Describing an event or a set of observations. Participating in brainstorming and hypothesizing before an investigation. Formulating initial and follow-up questions in the study of a topic. Summarizing data, interpreting findings, and offering conclusions. Questioning or refuting previous findings. Making decisions. Use of a map to show the locations of high- and low-pressure systems. Use of weather symbols common to meteorology.

Predicting: Making a forecast of future events or conditions in the context of previous observations and experiences.

> *Example:* Stating, "Based on previous data, the climate is gradually warming."

Science Process Skills (cont.)

Manipulating Materials: Handling or treating materials and equipment skillfully and effectively.

> *Example:* Using a wind vane to measure wind direction. Using a thermometer to measure temperature. Using a barometer to measure barometric pressure.

Using Numbers: Applying mathematical rules or formulas to calculate quantities or determine relationships from basic measurements.

> *Example:* Measuring and charting temperature, wind speed and direction, relative humidity, and barometric pressure.

Developing Vocabulary: Specialized terminology and unique uses of common words in relation to a given topic need to be identified and given meaning.

> *Example:* Using context clues, working definitions, glossaries or dictionaries, word structure (roots, prefixes, suffixes), and synonyms and antonyms to clarify meaning, i.e., cirrus, stratus, cumulus, altostratus, altocumulus, cumulonimbus.

Questioning: Questions serve to focus inquiry, determine prior knowledge, and establish purposes or expectations for an investigation. An active search for information is promoted when questions are used.

> *Example:* Using what is already known about a topic or concept to formulate questions for further investigation; hypothesizing and predicting prior to gathering data; or formulating questions as new information is acquired.

Using Clues: Key words and symbols convey significant meaning in messages. Organizational patterns facilitate comprehension of major ideas. Graphic features clarify textual information.

> *Example:* Listing or underlining words and phrases that carry the most important details, or relating key words together to express a main idea or concept.

INTEGRATED PROCESS SKILLS

Creating Models: Displaying information by means of graphic illustrations or other multisensory representations.

> *Example:* Drawing a graph or diagram; constructing a three-dimensional object, using a digital camera to record an event, constructing a chart or table, or producing a picture or map that illustrates information about current or future weather.

Formulating Hypotheses: Stating or constructing a statement that is testable about what is thought to be the expected outcome of an experiment (based on reasoning).

> *Example:* Making a statement to be used as the basis for an experiment: "If the barometric pressure changes, a change in the weather will occur."

Science Process Skills (cont.)

Generalizing: Drawing general conclusions from particulars.

> *Example:* Making a summary statement following analysis of experimental results: "The overall temperature of an area increases as the rays of the sun strike the area at a more direct angle."

Identifying and Controlling Variables: Recognizing the characteristics of objects or factors in events that are constant or change under different conditions and that can affect an experimental outcome, keeping most variables constant while manipulating only one variable.

> *Example:* Taking and recording weather readings at the same time of day and in the same location.

Defining Operationally: Stating how to measure a variable in an experiment; defining a variable according to the actions or operations to be performed on or with it.

> *Example:* Defining snowfall as the average of five different measurements of snowfall.

Recording and Interpreting Data: Collecting bits of information about objects and events that illustrate a specific situation, organizing and analyzing data that has been obtained, and drawing conclusions from it by determining apparent patterns or relationships in the data.

> *Example:* Recording data (taking notes, making lists/outlines, recording numbers on charts/graphs, making tape recordings, taking photographs, writing numbers of results of observations/measurements) from observations to determine an overall view of the current weather conditions.

Making Decisions: Identifying alternatives and choosing a course of action from among alternatives after basing the judgment for the selection on justifiable reasons.

> *Example:* Determining optimum location(s) for weather data collecting station(s).

Experimenting: Being able to conduct an experiment, including asking an appropriate question, stating a hypothesis, identifying and controlling variables, operationally defining those variables, designing a "fair" experiment, and interpreting the results of an experiment.

> *Example:* Formulating a researchable question, identifying and controlling variables including a manipulated and responding variable, data collection, data analysis, drawing conclusions, and formulating new questions as a result of the conclusions.

Definitions of Terms

Air pressure is a measurement of force exerted on a given unit of space by the weight of air; it is also known as barometric pressure.

Angle of incidence refers to the angle at which the sun's rays strike the earth's surface.

Anticyclones form in high-pressure areas with cold, dry air that spiral out in a direction opposite to cyclones, usually bringing clear, dry, and fair weather.

Atmosphere is the air that surrounds the earth; the majority of our weather data is taken from the troposphere, the layer of the atmosphere closest to the earth, which extends up seven to eight miles.

Autumn begins on September 22 or 23.

A **barometer** is an instrument used for measuring air pressure.

A **blizzard** results when wind speeds exceed 56 km/h and the temperature is less than -7°C.

The temperatures and precipitation for a particular area over a long period of time is called the area's **climate**.

Cirrus clouds form at high altitudes. They are long and feathery and made of ice crystals. They are often seen before a weather change, such as rain or snow.

Cloud droplets are tiny drops of water that form as water condenses, thereby forming clouds.

Clouds are masses of water droplets or ice crystals located in the troposphere layer of the atmosphere.

If a cold air mass overtakes a warm air mass, a **cold front** forms.

Condensation is a physical phase change from the property of gaseous matter to liquid matter.

Condensation nuclei are tiny particles (e.g., smoke and dust) around which cloud droplets form.

Continental air masses build over land.

Dryer **continental climates** exist within large landmasses. Temperatures swing more wildly from high to low. Summers may be hot, and winters quite cold.

Continental polar air masses form over northern Canada and may cause the mercury in thermometers in northern states to nose-dive.

Continental tropical air masses occur in the summer and form over Mexico, bringing hot air to the southwestern states.

Condensing water vapor from the exhaust of jet engines forms condensation trails, often called **contrails**.

The **Coriolis Effect** is the deflection of winds caused by the rotation of the earth on its axis.

Cumulus clouds usually form at mid-altitudes. These flat-bottomed, fluffy clouds are associated with fair weather.

Spinning air masses called **cyclones** form when cool air swoops in to replace rising warm air in a region of low pressure.

Definitions of Terms (cont.)

Closer to the earth, water vapor condenses on a surface, such as grass, and can be observed in the form of **dew** or frost, if the temperature is below freezing.

Dew point is a measure of humidity in the air (e.g., the temperature at which dew will begin forming).

The **doldrums** are mostly very calm air in a band over the equator.

Elevation, or altitude, of an area is its distance above sea level.

Evaporation is the physical change from the property of liquid matter to gaseous matter.

Air is considered to be a **fluid** as it takes the shape of its container and can flow.

Fog is a cloud that forms at the surface of the earth.

Points of contact between air masses are called **fronts**.

Global winds are winds that blow over long distances from specific directions.

Water can also be contained in the ground in the form of **groundwater**. This groundwater may be hidden in underground aquifers or in-ground.

Heat refers to a form of energy that flows from one object to a cooler object. Equal amounts of heat applied to two objects will not necessarily result in the same temperature difference; it depends upon the material.

Regions of sinking cool air are called **high-pressure systems**, or anticyclones.

Humidity is the moisture in the air.

A **hurricane** is a tropical cyclone with wind speeds of at least 74 mph or more.

A **hygrometer** was the first instrument designed for measuring humidity.

Isobars are lines on a weather map that connect areas of equal barometric pressure.

Isotherms are lines on a weather map that connect areas of equal temperature.

Latent heat is energy stored when evaporation turns liquid into a gas and is later released when condensation occurs.

Latitude is the distance of a place north or south of the equator.

Local winds are winds that blow over short distances.

Regions of rising warm moist air are called **low-pressure systems**, depressions, or cyclones.

The large bodies of water around **marine climates** give up moisture to winds, providing more precipitation. Summers tend to be warm (rather than hot) and winters mild.

Maritime air masses assemble over oceans.

Maritime polar air masses form over the Pacific Ocean and the North Atlantic. They carry cool moist air.

Definitions of Terms (cont.)

Maritime tropical air masses form over the ocean near the equator and can bring hot, humid summers or stormy winters if they bang into cold, northerly air heading south.

Scientists who study the earth's atmosphere are called **meteorologists**.

Meteorology is the study of weather, and it deals with understanding the forces and causes of weather.

Mountain ranges alter the flow of prevailing winds.

The **National Weather Service** was established in 1870 by the United States government to keep people informed of changing weather conditions.

When two cold air masses collide and push a warm air mass up between them, an **occluded front** forms.

At night, the water stays warm after the land has cooled, resulting in an **offshore breeze**.

Polar, or **arctic, climate zones** exist between latitudes of 90° and 60°. Here, near either pole, there is no summer, and the average yearly temperature remains below 0°C.

Polar easterlies are cold, fairly weak winds blowing from east to west.

Precipitation occurs when water in liquid and solid form can no longer stay in the earth's atmosphere.

Prevailing westerlies are cool air, usually moving quickly toward the poles from west to east in both hemispheres.

Prevailing winds are winds that blow more often from one direction than the other.

Radar (**ra**dio **d**etecting **a**nd **r**anging) instruments send out radio signals that are reflected from objects such as clouds and rain.

Rainstorms and snowstorms result from the collision of different fronts. Nimbostratus clouds form when a warm front moves in and rises over cold air, often resulting in heavy rain or snow.

Relative humidity refers to the amount of moisture in a given amount of air relative to what could be contained if the given amount of air were completely saturated.

Cooler moist air from over the water slides beneath the rising air, creating a **sea breeze** on an ocean beach.

Severe weather is weather that can potentially cause harm to people and the environment. Hurricanes, tornadoes, thunderstorms, and blizzards are all examples of potentially dangerous storms.

Spring in the Northern Hemisphere begins on March 20 or 21.

A **stationary front** occurs when air masses tend to remain in place for a period of time.

Stratus clouds form at low altitudes. They form layers of sheet-like clouds and are associated with steady rains, snow, and very cloudy days.

Definitions of Terms (cont.)

Temperate zones extend from latitudes 60° to 30°. Here there are seasonal swings in temperature, and the average temperatures can vary from 5° to 20°C.

Temperature is the average kinetic energy (of the particles) of an object.

The **temperature of ocean currents** directly affects the temperature of the air above them.

A **thermoscope** is the earliest form of a thermometer.

Thunderstorms form when a cold front moves in and meets a warm front.

Tornadoes are very destructive storms that form over land with destructive winds that can spin up to 300 mph (480 kp/h).

Trade winds are warm, steady winds that blow back toward the equator in usually clear skies.

The **tropical zone** bands the earth from the equator at 0° to 30° on either side of the equator. Temperatures are hot, and there is high humidity.

The **troposphere** is the layer of the atmosphere closest to the earth. Seventy-five percent of the atmosphere's gases are located here, and all weather occurs in this layer.

Storms that form over the western Pacific Ocean are called **typhoons**.

If a warm air mass overtakes a cold air mass, a **warm front** is created.

The **water cycle** is the exchange of water between land, bodies of water, and the atmosphere. It is also known as the hydrologic cycle.

Water spouts form over water. This storm may cause great damage with winds that spin up to 95 km/h.

Water vapor is water in a gaseous state; water vapor is invisible.

Weather is the condition of the atmosphere at a particular time and place in a region.

Weather boxes are simply an organized collection of symbols used to represent weather data.

Weather satellites, or meteorological satellites, are orbiting high above Earth's surface in the atmosphere.

Wind, air in motion, is a result of the uneven heating of the earth's surface by the sun.

High, cumulonimbus clouds produce thunder, lightning, and sudden air movements called **wind shears,** which are dangerous for planes.

Answer Keys

Historical Perspective
Quick Check (page 6)
Matching
1. c   2. d   3. b   4. a   5. e
Fill in the Blank
6. meteorology
7. Galileo Galilei
8. Daniel Gabriel Fahrenheit
9. Benjamin Franklin
10. National Weather Service
Multiple Choice
11. d   12. b   13. c   14. a

Weather
Quick Check (page 10)
Matching
1. c   2. a   3. e   4. b   5. d
Fill in the Blank
6. weather   7. sun
8. pressure   9. high pressure, low pressure
10. Coriolis Effect
Multiple Choice
11. b   12. c   13. a

The Sun's Effect on the Atmosphere
Quick Check (page 14)
Matching
1. a   2. d   3. b   4. e   5. c
Fill in the Blank
6. radiant energy   7. rotation, revolves
8. atmosphere   9. sun
10. equator
Multiple Choice
11. b   12. a   13. b
Knowledge Builder (pages 15–16)
Activity #1:
Conclusion: Answers may vary but should include the angle at which light from the sun strikes the earth results in uneven heating of the earth's surface. The more direct rays cause greater heating.
Activity #2:
Conclusion: Answers may vary but should include that the sharper the angle at which the sunlight strikes the earth, the more area that light covers. The same radiant energy is spread across more area.
Activity #3:
1. Answers may vary but should include the earth's orbit is not circular. Earth's orbit is elliptical, so its distance from the sun varies.
2. Answers may vary but should include the sun's rays have to travel a longer distance before they reach the poles because of the earth's tilt. The poles are either tilted toward the sun or away from the sun at different

times of the year. The equator is closest to the sun throughout the year. As a result, the equator receives the most solar energy.
3. Answers may vary but should include there would be no seasons. The seasons are a result of the sun's light reaching the earth at a tilted angle.
4. Answers may vary but should include locations on Earth would have six months of darkness and six months of light. With such extremely long days, the sunny side would have extremely hot temperatures while the shadowed side would have extremely cold temperatures. All locations would experience extreme heat and extreme cold during the course of a year.
Inquiry Investigation (pages 17–18)
Hypothesis: The color of a surface will/will not affect temperature.
Conclusion: Answers may vary but should include a brief description of what happened in the experiment and whether or not the hypothesis was supported by the data collected. Students should recognize that there is a variation in temperature according to the color of the surface.

Air Pressure
Quick Check (page 21)
Matching
1. e   2. d   3. a   4. c   5. b
Fill in the Blank
6. weight   7. low-pressure system
8. fair weather   9. stormy weather   10. L, H
Multiple Choice
11. a   12. b   13. c
Knowledge Builder (page 22)
Activity #1:
The egg fell into the bottle.
Conclusion: Answers may vary but should include hot air expands. Cold air contracts. When the air inside the bottle is heated, the molecules, or tiny air particles, inside the bottle spread out, increasing air pressure. As the air in the bottle cools, the air pressure decreases. The greater outside air pressure pushes the egg into the bottle. Blowing into the bottle raises the air pressure again. The air and the egg rush out of the bottle.
Activity #2:
Conclusion: Falling barometric pressure indicates stormy weather. Rising barometric pressures indicates fair weather.

Activity #3:

1. The lines should indicate barometric pressure decreases as one moves to the center of the low-pressure area and increases as one moves to the center of a high-pressure area.
   A. 1114 mb   B. 1118 mb
   C. 1122 mb   D. 1038 mb
2. low-pressure area in upper Midwest and high-pressure area in the west-Southwest
3. low-pressure area in Michigan; high-pressure area in West-Southwest
4. Michigan: stormy weather; West-Southwest: fair weather

The Water Cycle
Quick Check (page 27)
Matching

1. c   2. d   3. e   4. a   5. b

Fill in the Blank

6. Condensation   7. clouds   8. sun
9. precipitation   10. evaporation

Multiple Choice

11. c   12. a   13. d

Knowledge Builder (page 28)
Activity #2:

Observation: Water formed on the inside walls of the bag.

Conclusion: Answers may vary but should include heat from the sun causes the water to evaporate and become a vapor. As the water vapor cools, it condenses, forming tiny droplets which gather to form clouds. As the droplets get larger, they become heavier, causing them to fall to the ground as precipitation.

Wind
Quick Check (page 31)
Matching

1. d   2. c   3. a   4. e   5. b

Fill in the Blank

6. Solar energy   7. poles, equator
8. offshore breeze   9. sea breeze
10. direction

Multiple Choice

11. d   12. a   13. b

Knowledge Builder (page 32)
Activity #2:

Conclusion: Answers may vary but should include it helps meteorologists identify weather patterns.

Clouds
Quick Check (page 36)
Matching

1. d   2. e   3. a   4. b   5. c

Fill in the Blank

6. atmospheric   7. 100   8. humidity
9. Relative humidity   10. Stratus, cumulus, cirrus

Multiple Choice

11. b   12. b   13. c   14. b

Knowledge Builder (pages 37–38)
Activity #1:

Observation: Answers will vary.

Conclusion: Answers may vary but should include as the air outside the can cooled, the water vapor in the air condensed. Condensed water appears as tiny water droplets on the side of the can.

Activity #2:

Conclusion: Answers may vary but should include water vapor condenses into the form of small cloud droplets. Adding particles, such as the smoke, enhances the process of water condensation. Water particles will group together more easily if there are some solid particles in the air to act as nuclei. Squeezing the bottle causes the air pressure to drop. This creates a cloud.

Activity #3:

Observation: Answers may vary but should include different kinds of clouds indicate different kinds of weather.

Conclusion: Answers may vary but should include meteorologists can use clouds to help forecast weather because different kinds of clouds indicate different kinds of weather.

Activity #4:

Conclusion: Answers may vary but should include air at different temperatures is capable of holding varying amounts of moisture.

Severe Weather
Quick Check (page 42)
Matching

1. d   2. e   3. b   4. a   5. c

Fill in the Blank

6. tropical cyclones   7. Wind speeds
8. Tornado Alley   9. Hurricanes, tornadoes
10. Severe weather

Multiple Choice

11. a   12. b   13. c   14. a

Knowledge Builder (page 43)
Activity #2:

Observation: The water formed a funnel. The direction the funnel swirls depends on the direction the student swirled the water.

Mapping Weather
Quick Check (page 46)
Matching
1. c   2. d   3. b   4. e   5. a
Fill in the Blank
6. Weather data   7. weather symbols
8. Weather boxes   9. radar images
10. atmosphere
Multiple Choice
11. d   12. a   13. d

Knowledge Builder (pages 47–48)
1. 28°F
2. 15–20 mph or 15 knots
3. hurricane
4. west
5. 30.6 in
6. falling, symbol used on map, other states around Missouri have the same reading
7. 75%
8. drizzle
9. Florida
10. stormy weather headed to Tampa because hurricane in Miami, winds blowing from southeast.
11. a.   b.

c.   d.

Weather Fronts
Quick Check (page 52)
Matching
1. c   2. e   3. a   4. b   5. d
Fill in the Blank
6. stationary front   7. Maritime polar
8. cold front   9. tornadoes
10. Maritime tropical
Multiple Choice
11. c   12. a   13. c

Knowledge Builder (page 53)
1. A cold front is moving in from the northwest to the southeast. A cold front usually moves from the northwest to the southeast.
2. A cold front symbol should be used from Michigan to Arkansas.
3. A heavy rain shower can be expected.

Climate
Quick Check (page 58)
Matching
1. d   2. c   3. b   4. e   5. a
Fill in the Blank
6. Mountain ranges   7. ocean currents
8. marine, continental climates
9. Prevailing winds   10. climate zones
Multiple Choice
11. d   12. b   13. a

Bibliography

Weather Book List:

Allaby, Michael. *Blizzards*. New York: Facts On File. 2003

Allaby, Michael. *How Weather Works*. London: Dorling Kindersley. 1999.

Allen, Jean. *Tornadoes*. Mankato, MN: Capstone Books. 2000.

Armbruster, Ann and Elizabeth A. Taylor. *Tornadoes*. London: Franklin Watts. 1993.

Arnold, Caroline. *El Niño*. New York: Clarion Books. 1998.

Bendick, Jeanne. *The Wind*. Skokie, IL: Rand McNally and Co. 2000.

Berger, Melvin and Gilda Berger. *Do Tornadoes Really Twist?* New York: Scholastic Reference. 2001.

Bower, Miranda. *Experiment With Weather*. Minneapolis, MN: Lerner Publications. 1993

Bramwell, Martyn. *Weather*. London: Franklin Watts, 1994.

Crowder, Bob, Ted Robertson, and Elinor Vallier-Talbot. *Weather*. Sydney: Time-Life Books. 1996.

Burton, Jane and Kim Taylor. *The Nature of and Science of Rain*. Strongsville, OH: Gareth Stevens Publishing. 1997

Challoner, Jack. *Hurricane and Tornado*. London: Dorling Kindersley. 2000.

Christian, Spencer. *Can It Really Rain Frogs?* Hoboken, NJ: John Wiley and Sons. 1997.

Clark, John. *The Atmosphere*. London: Franklin Watts. 2000.

Cosgrove, Brian. *Weather*. London: Dorling Kindersley. 2002.

Dickinson, Terence. *Exploring the Sky by Day: The Equinox Guide to Weather and the Atmosphere*. Ontario: Firefly Books. 1988.

Dineen, Jacqueline. *Hurricanes and Typhoons*. London: Franklin Watts. 1995.

Eden, Philip and Clint Twist. *Weather Facts*. London: Dorling Kindersley. 1995.

Farndon, John. *Weather*. London: Dorling Kindersley. 1992.

Flint, David. *Weather and Climate*. London: Franklin Watts. 1991.

Fradin, Dennis B. *Blizzard, Disaster!* New York: Children's Press. 1985.

Galiano, Dean. *Clouds, Rain, and Snow*. New York: Rosen Publishing. 2000.

Galiano, Dean. *Hurricanes*. New York: Rosen Publishing. 1999.

Galiano, Dean. *Thunderstorms and Lightning*. New York: Rosen Publishing. 2003.

Galiano, Dean. *Tornadoes*. New York: Rosen Publishing. 2000.

Ganeri, Anita. *And Now—The Weather*. London: Chrysalis Children's Books. 1992.

Ganeri, Anita and Roger Hunt. *The Usborne Book of Weather Facts*. London: Usborne. 1987.

Gentle, Victor. *Hurricanes*. Strongville, OH: Gareth Stevens Publishing. 2001.

Graedel, T.E., and Paul J. Crutzen. *Atmosphere, Climate, and Change*. New York: Scientific American Library. 1995.

Haslam, Andrew and Barbara Taylor. *Weather: Make It Work!* Lanham, MD: Cooper Square Publishing. 2000.

Bibliography (cont.)

Humphrey, Paul. *The Weather*. London: Franklin Watts. 1996.

Kahl, Jonathan. *National Audubon Society First Field Guide: Weather*. New York: Scholastic. 1998.

Kent, Deborah. *Benjamin Franklin: Extraordinary Patriot*. New York: Scholastic. 1993.

Lampton, Christopher. *Tornado*. Minneapolis, MN: Milbrook Press. 1991.

Llewellyn, Claire. *Wild, Wet and Windy*. Somersville, MA: Candlewick Press. 1998.

Ludlum, David. *National Audubon Society Field Guide To North American Weather*. New York: Alfred A. Knopf. 1991.

Lyons, Walter. *The Handy Weather Answer Book*. Canton, MI: Visible Ink. 1991.

Mason, John. *Winter Weather*. London: Hodder. 1990.

McMillan, Bruce. *The Weather Sky*. New York: Farrar Straus Giroux. 1996.

Merk, Ann and Jim Merk. *Storms, Weather Report*. Vero Beach, FL: Rourke Publishing. 1994.

Merk, Ann and Jim Merk. *Weather Signs*. Vero Beach, FL: Rourke Publishing. 1994.

Mogil, H. Michael. *Weather, An Explore Your World Handbook*. Ludlow, UK: Discovery Books. 1999.

Morgan, Sally. *Weather*. London: Hodder. 1993.

Murphee, Tom and Mary K. Miller. *Watching Weather*. Brighton, Victoria, Australia: Owl Publishing. 1998.

Oard, Michael. *The Weather Book*. Green Forest, AR: Master Books. 1997.

Onish, Liane. *Wind and Weather*. New York: Scholastic. 1995.

Parker, Steve. *Weather: Fun With Science*. Boston, MA: Kingfisher Books. 1997.

Rotter, Charles. *Tornadoes*. Mankato, MN: Creative Education. 1997.

Ruiz, Andres Llamas. *Rain*. New York: Sterling Publishing Co. 1997.

Silverstein, Alvin, Virginia Silverstein, and Laura Nunn. *Weather and Climate*. Brookfield, CT: 21st Century Books. 2007.

Simon, Seymour. *Tornadoes*. New York: Harper Collins. 2001.

Svarney, Patricia and Thomas Svarney. *Skies Of Fury*. Austin, TX: Touchstone. 1999.

Tannenbaum, Beulah. *Making and Using Your Own Weather Station*. State College, PA: Venture Books. 1989.

Taylor, Barbara. *Wind and Weather*. London: Franklin Watts. 1991.

Tripp, Nathaniel. *Thunderstorm!*. New York: Dial Books. 1994.

Tufty, Barbara. *1001 Questions Answered About Hurricanes, Tornadoes and Other Natural Air Disasters*. Mineola, NY: Dover. 1987.

Van Rose, Susanna. *The Earth Atlas*. London: DK Children. 1994.

Walker, Jane. *Air: Against The Elements*. London: Franklin Watts. 2000.

Williams, Jack. *The Weather Book*. New York: Vintage Books. 1997.

Wood, Jenny. *Storms: Facts, Stories, Activities*. Lanham, MD: Cooper Square Publishing. 1999.

Bibliography (cont.)

Weather Curriculum Materials:

BSCS Science T.R.A.C.S. Series
Investigating Weather, 2, Teacher's Edition
Investigating Weather Systems, 5, Teacher's Edition
Investigating Weather Systems, 5

DSM II Series, Delta Education, Teacher's Guides
Amazing Air, 2–3
Weather Instruments, 3–5
Solar Energy, 5–6
Weather Forecasting, 5–6

Overhead and Underfoot, 3–5, AIMS Education Foundation, Revised Edition

Project Atmosphere, American Meteorological Society
Clouds
Hazardous Weather
Jet Streams
Water Vapor and the Water Cycle
Today's Weather
Weather Satellites

Project Earth Science: Meteorology (2nd Edition). 2001. National Science Teachers Association. Arlington, VA.

Project SafeSide, On The Safe Side With The Weather Channel

Wild About Weather, National Wildlife Federation, Learning Triangle Press

Recommended Websites:

www.education.noaa.gov
www.usatoday.com
www.weather.com
www.weatherworks.com
www.weatherdesk.org

Bibliography/Reference:

Brunet, C., Holle, R., Mogil, H., Moran, J., Phillips, D. *Weather*. New York: Discovery Books. 1999.

Crowder, Bob, Ted Robertson, and Elinor Vallier-Talbot. *Weather*. Sydney: Time-Life Books. 1996.

Eden, Philip and Clint Twist. *Weather Facts*. London: Dorling Kindersley. 1995.

Graedel, T.E., and Paul J. Crutzen. *Atmosphere, Climate, and Change*. New York: Scientific American Library. 1995.

Murphee, Tom and Mary K. Miller. *Watching Weather*. Brighton, Victoria, Australia: Owl Publishing. 1998.

Sager, Robert, William Ramsey, Clifford Phillips, and Frank Watenpugh. *Modern Earth Science*. Austin: Holt McDougal. 2002.

Williams, Jack. *The Weather Book*. New York: Vintage Books. 1997.